T0064534

VISIONS OF 2050

ROGER BOURKE WHITE JR.

authorHOUSE®

AuthorHouse™
1663 Liberty Drive
Bloomington, IN 47403
www.authorhouse.com
Phone: 1 (800) 839-8640

Published by AuthorHouse   08/31/2015

ISBN: 978-1-5049-3405-3 (sc)
ISBN: 978-1-5049-3406-0 (e)

Print information available on the last page.

Books by Roger Bourke White Jr.

Tales of Technofiction Series

Child Champs
The Honeycomb Comet
Rostov Rising
Science and Insight for Science Fiction Writing
Tips for Tailoring Spacetime Fabric, Vol. 1
Tips for Tailoring Spacetime Fabric, Vol. 2

Business and Insight Series

Goat Sacrificing in the 21st Century
Evolution and Thought: Why We Think the Way We Do
How Evolution Explains the Human Condition
Surfing the High Tech Wave: A History of Novell 1980–1990

Contents

Introduction

Welcome to *Visions of 2050*. This book is my speculations on what life will be like thirty-five years from now, when cyber is both self-aware and in full control of big business, and the large-scale manufacturing and service operations that are currently becoming more and more automated. What I envision in this book is that the automation of these areas will be completed. Self-aware cyber entities will handle making most of what humanity consumes, such as large-scale material goods manufacturing, large-scale services, and most motorized transportation.

The result of this change is a whole lot of fascinating changes in how humans live and what they think about. For instance, with the cyber handling of most manufacturing, transportation and service jobs, humans are going to have to find other kinds of employment. What will humans be doing instead? What will be different in what is taken for granted? What will be different in what is considered the important things in their lives? What are the things they will think about a lot?

A related change is the development of what I call cyber muses. These are cyber entities specifically designed to interact with humans. These are the technological extension of things such as computer games, hospitality bots, robocallers, and smart advertising placers. The "muse" moniker came to me because many of them will be designed to bring out the best in the humans they serve, but they will also serve many other purposes. (Discovering these others is the fun part of exploring ramifications.)

In addition, this book addresses the ramifications of the many other technological advances that are coming. Again, the ramifications mean talking about how they will change how we live and think. What will we think and worry about, and what will we take for granted? This is the main function of the stories in this book. They explore human thinking.

Outline and Definitions

Outline

The most important difference between the 2050s and the 2010s is that cyber (my term for the pervasive network of computing power that will be installed by then) will be running the large-scale manufacturing, service, and transportation industries of the world, which I will refer to as "big business". What follows in this book are the ramifications of that.

The chapters are divided into two categories: nonfiction essays about the specifics of what will be happening in 2050 and science fiction stories that will highlight the ramifications of these changes. The latter are stories about how these changes will affect how we humans live and think in the 2050s environment.

The layout of this book is not standard. This is a topic that doesn't fit easily in standard nonfiction or fiction formats. The definition section that follows is an example of this being a nonstandard format. The definitions both define and begin the discussions on these various topics.

Shorter Definitions

Big Business: As discussed above, big business is those manufacturing, service, and transportation activities taking place on a large scale that have become fully automated. An example is that the choice of what products to make in a big business factory will be made by cyber in charge of the factory talking to cyber in other members of the supply chains it is part of instead of by human managers. These organizations are not just businesses. They can be guilds and government agencies as well—think of what happens to regional bus authorities and taxi guilds as buses and taxis become driverless. Operating buses and taxis becomes part of big business as it is defined in this book.

The Curse of Being Important: This is my version of the proverb about too many cooks spoiling the broth. I write about it a lot in my books on human thinking (my *Business and Insight* series). When a project is afflicted with The Curse, progress slows dramatically because so many people have important opinions that affect how the project progresses.

Hobbiton: A neighborhood layout that emphasizes being cute and comfortable rather than efficient as regards to traffic and unpleasant transportation surprises such as snow storms, floods, and routine roadwork. Hobbiton-style means that the roads will dodge structures that are culturally important, include lots of cul-de-sacs, emphasize walkability and bikeability over driveability, and include many hard-to-change historic districts and gated communities. All-in-all, it is a style that will be fun to walk and bike around, but it will be killer for getting things done quickly and efficiently when cars and trucks are involved.

Sacred Feminine and Sacred Masculine: The sacred feminine gets a lot of coverage in fiction. It is the idea that the world would be a better place if feminine ideals such as gender equality and world peace were more widely implemented. I add to this the concept of the Sacred Masculine. This concept is closely linked to enfranchisement. The goal of keeping the sacred masculine in mind is to combat a really old instinct, a prehuman one. This is the male loner instinct that is widespread among mammalian species. In many species, adult males are loners except during mating season when they gather to fight over access to the females that are in heat. When this instinct is allowed to flower, the males of the community will not really give or take much from the community. The current stereotype of this is the male who spends all his time in his parents' basement playing video games. To prevent having this behavior become widespread, the sacred masculine must be recognized as a valid concept and stroked just as rigorously as the sacred feminine. Men must be put on a recognition pedestal just as much—or more so—than women.

Cyber names in stories: When cyber has a name, it will be a name plus a number, such as Nancy-123. I will use this full name once and then typically shorten it to just Nancy.

"This is unclear": Some ramifications are easier to infer than others. When I use this phrase, it means that there is still a whole of lot fuzziness in my mind about the ramifications of the topic being discussed.

Zombie mode: This is when people are paying lots of attention to their internal communications systems while they walk or do other activities. It is like walking while phoning is today—but with better technology.

Harsh reality: This is the physical world we actually live in. It doesn't have to be harsh, but I call it that because it is referring to the world of hard limits on what we can do and accomplish. It is the converse to what we are deeply wishing and hoping will happen.

Exotic products: There are situations in some of these stories that require commerce to make sense, as in, enough trading of stuff that an industry can be created. When commerce is needed I come up with a fictional substance or product that is quite valuable to justify the commerce. Examples being Yalmamma in the story I am a Rock, and Pencrock in Youth is Wasted on the Young ... Hah!

Ground-pounder: A military slang term referring to soldiers who have to walk from place to place. In the 20th century these were the infantry. In this story it is people living on Earth who are Earth-centric, they don't think much about what is happening in the rest of the solar system.

Longer Definitions and Axioms

These are concepts that take more to explain and define.

Humans and Cyber

Cyber are the intelligent, self-aware cyber entities that will inhabit cyber space. By 2050, there will be a lot of cyber space and a lot of self-aware entities of many different kinds inhabiting it. Like cyberspace itself, the number will grow exponentially in number and sophistication. Cyber beings will create their own new species, with each generation being more sophisticated than the previous. Most of these will be too intelligent for humans to understand or even grasp that they exist and what they do, but there will still be many of these entities that interact with humans. Cyber and humans will cooperate on getting human-understandable projects designed and completed, both in cyberspace and the tangible world.

In this series, I will be concerned with only cyber that interacts meaningfully with humans. Those that will be too advanced for humans to comprehend are too advanced for me to be writing about.

The cyber entity forms will vary enormously. Some will remain purely cyber, and they never leave cyberspace. They will interact with humans purely through conversations of various sorts. These conversations may be through voice or sight. They might also occur by adjusting the numerous wearables that humans will have, or by adjusting the manufacturing and service industries that produce goods and services for the world. I call these latter designer economic booms and recessions.

Others will interact by temporarily inhabiting avatars of various sorts. They will be mostly cyber, but they can become temporal when there is compelling reason to do so. When the pure cyber types inhabit avatars, they will be clumsy at it. This will be a signal to others that they are indeed pure cyber types.

Still, others will be creations. They will spend most of their time in a physical form and routinely interact with humans.

Can cyber die, as in lose its self-awareness in some fashion and not be replaceable? Given the ease of making backups, this is hard to imagine, even in those forms that are creations. This is one attribute of cyber that will make them distinctive from humans and organic-based androids–the kind that are grown from an immature form like humans are. If cyber beings do die, it is because other cyber have killed them. An example of this is eliminating bad cyber such as viruses, worms and other malware. Another reason for it happening is to clear out cyberspace so there is room for more evolved versions. This is essentially the same reason that evolution favored mortality for higher physical forms of organic life.

Creations, Androids, and Avatars

Creations, androids, and avatars are the tangible forms of cyber. These are the forms with bodies in the real world. First, some definitions.

- *Creations.* Self-aware, mechanical-based beings. These can range from being very much like machines, such as satellites exploring space, to being very humanlike, such as tour guides. Most of their thinking will happen within the tangible being. Some of their intelligence can be in cyberspace, and most will be closely linked with cyberspace, no matter where the core of their intelligence resides. Like pure cyber, they can be backed up and replicated. They can also be reprogrammed with upgraded or completely new intelligence.

- *Androids.* Self-aware and organic, but artificially grown. These can be quite human in appearance and nature, but they are considered different from humans. They can also depart widely from the human design. They can be designed to live in environments too extreme for humans, such as the low gravity of spaceships and moons around the solar system. They can also be designed as tradeoffs (for example, extremely high-performance, but short-lived). They can be grown from immature forms as humans are or assembled in vats to come out fully formed and functional. This latter will tend to be used for special purposes and short-lived. When innovation and adaptability are important, these creatures will be grown much as humans are, from immature fetuses who learn as they grow.

- *Avatars.* These are mechanical or androidlike in form but are not self-aware. These are designed to be remotely controlled, so they are tightly linked to cyberspace. They aren't considered beings because they aren't self-aware. They can be controlled by humans or cyber.

Cyber Muses

A surprise use of cyber and androids will be to become inspiring to humans. "Behind every great man is a good woman." is the proverb behind this. Cyber muses are a case where cyber is designed to replace the good woman as an inspiration source and to provide it faster, better, and cheaper.

These are cyber beings designed specifically to inspire humans to great accomplishments in various activities and to stroke emotions that humans wish to have stroked. Note that these are two very different tasks. The emotion stroking will be easier to design and come first. For example, stroking to feel great sex is likely to be one of the first followed quickly by a cute kittens-style of comfort stroking. Sometime later will come the classic muse activities of inspiring the creating of great art and great leadership. Many other kinds of inspiring and stroking will be developed as well. These will include things such as inspiring engineering, inventing, politicking, and stroking keeping-up-with-the-Joneses emotions. This final kind I call arm-candy cyber muses.

Colonizing Moons and Asteroids

This is man and cyber heading into space. This will be establishing bases on the moon, Mars, Mercury, asteroids, and the moons of the gas giants—the less extreme environments in the solar system. (The more extreme are places like the surfaces of the gas giants.) At first, these bases will be small and cyber-manned, and they will stay that way for a long, long time.

What will change these bases from small and scientific to something bigger and more diverse is finding good answers to the question of what is commercially exploitable. As long as the answer is nothing, which is the current answer, this activity will remain small-scale and almost completely science-oriented, with a slight bit of billionaire-oriented space tourism mixed in.

If something on these planetary bodies could be transformed in a profitable way, the activity will become larger and more diverse. It will be larger and more diverse in direct proportion to what can be commercially profitable. Think of the huge profits of the spice trade that powered the emergence of the ship-borne trading industry from Western Europe to the Far East in the 1600s and 1700s. This industry emerged when it did because advancing ship technology finally allowed huge profits to be made.

The only other reason to be out there is to sustain "crazy people", such as the early colonists of New England were—the Puritans and Quakers. ("crazy" in this context means belief that is not conventional for its time and place) But keep in mind that even though they went over for crazy reasons (religious freedom), they stayed and thrived because they found commercially successful activities to engage in. The Vikings (and perhaps the Irish) who crossed the Atlantic hundreds of years earlier never solved the commercially successful puzzle so they withered, as did their place in transatlantic history.

Very little in this book will be about interstellar travel. It is too expensive to be real in this time frame. If anything interstellar happens, it will be about interstellar-traveling aliens coming to the solar system. (I gave in to the fun side: I do end this book with a story about that happening.)

Further Reading

This November 2014 *Science News* article, "Rigors of Mars trip make teamwork a priority," by Bruce Bower is a lengthy article on how important physiological and psychological elements are to selecting good astronauts for lengthy missions (think months or even years). One of the common hazards is an inability to sleep well in space. Another is not staying part of the team by getting and acting isolated from the others.

Another point brought up by the article is the isolation from mission control. A ship in the vicinity of Mars will have a forty-minute turnaround on messages sent to Earth. This will affect cyber even more than humans because Earth-based cyber are inherently designed for gigabyte-speed interconnections. Since carrying cyber capabilities will be constrained by weight limits just as carrying humans are, the cyber in space will be of limited capacity and feel just as lonely as humans. Their capabilities in spaceships will be quite limited compared to what they are on Earth. This means two things. The first is that their communication skills will be primitive—ship cyber won't be able to stroke humans in the sensitive ways Earth cyber can. The second is spookier: Some form of the HAL of *2001: A Space Odyssey* may in fact be a real hazard for space ship cyber.

Creating Total Entitlement States (TES)

Most people in 2050 will not be working for a living in the current sense. Cyber will be handling what we call work today. Think of total factory automation, filling out forms, and driverless cars becoming ubiquitous taxis for all. These stories are about what humans will be doing instead these common jobs of today.

These new kinds of jobs will be dominated by the philosophy of doing what you want to do and what you have a passion for. The range will be large in potential, but like entertainment is today, it is likely to be dominated by a top-forty list of activities. These will be mostly emotion-driven choices. Only a few people will get into alternative, counterculture, or interdisciplinary employment outside the top-forty realm.

Trying to figure out the top-forty list is a challenge, but it is not too difficult because it will be so emotion-based. Thanks to total entitlement states (TES), people will not have to work for a living. Rather, they will get all their basic necessities through the TES system. Their work will be directed by their passions and fears.

Conversely, trying to figure out jobs that humans will do that will actually add value to society in the sense of increasing productivity and developing new disruptive technologies is going to be much tougher. Most people of this upcoming era will think of value-adding as doing some style of what is today considered artisanal manufacturing or service. The value addition is simply adding human sweat, tears, and mythology to the process. It is not developing ways of making things more productively. When the average human thinks about this issue, they will think that that is something cyber is responsible for.

A historical example of how surprising future productive jobs can be is as follows. A person who got into the delivery business in 1920 did so by learning how to drive a team of horses pulling a wagon and how to care for them when the work day ended. Their children, who got into the business in 1940, did so by learning how to drive a motor-powered truck and how to repair it when the work day ended. This is quite a dramatic transition in work skills. And the transition would have been quite a surprise to the dad. He would have been quite wrong on the advice he was giving his son who wanted to carry on the family business.

I'm facing forecasting this kind of transition in these stories.

Outlaw Activities

How will crime and violence, such as robbing, gangster, and terrorist activities, fit into this mix of what people do for work in the TES environment? In theory, the system can support them. The cyber may make such activities possible within certain ranges. Those people who engage in this will feel as though they are gaming the system—outlaws—and this will power their feeling of satisfaction. But the cyber know what is up and have figured out how to compensate for the damage they do to others with their outlaw behavior. When the violence does cross the line and becomes more damaging than cyber can compensate for, the pervasive surveillance will pick up the act and the perpetrators within minutes.

This brings up the question of what will going to jail be like. It won't be like it is today. Today's version is too expensive and doesn't solve the problem of successfully rehabilitating people. But the future solution will stroke similar "exile" emotions—the root instinctive emotion behind having jails. One of the related questions is how getting back into the community will be handled. How can the ex-con factor be handled better?

Money

One consequence of the change in employment practices will be that money as we know it today will no longer exist. The uniformity of today's money will be replaced by specialty styles of money. I envision three styles which I call necessity money, luxury money, and investment money. Note that having different styles of money is not new. One example of this fragmenting happening in the 2010s is cyber currencies such as Bitcoin.

Necessity money: As TES becomes established, this kind of money will buy the basic essentials of TES living—think of food stamps and their payment-card replacements.

Luxury money: This will buy luxuries. One of the interesting questions for the people living in 2050 is how to acquire this style of money. Artisanal-style human-crafting factories making luxuries will deal in this money.

Investment money: This is the money of big business. Because humans will have so little involvement with making big business happen, cyber will do the vast majority of the dealing with this style of money.

These different money styles will not interchange easily because their uses are so different. There will be black market dealings, but because cyber will be so involved in all money transactions, black markets will be small.

Wearables

Wearables are smart clothing and other smart devices that people put on themselves or into themselves. Wearables are going to become part of everyday living. Man! What won't wearables be involved in! Most will be taken for granted, except where envy is involved. The stories will be about the surprising implications. In the 2010s, the high-profile issue swirling around wearables is the tradeoff between privacy and the fast response to problems that intensive surveillance allows. This will be less so in 2050. By then, the value in lowering repair costs and raising convenience will be taken for granted, so the surveillance issue will diminish.

Accommodating Instinctive Human Worries

Even though people will have everything (at the basic-need levels), they will still come up with things to worry about. This is already widely demonstrated in the 2010s. Look at the popularity of if-it-bleeds-it-leads news stories in traditional and social media. In 2050, humans will still be finding lots to worry about.

What cyber can do is direct those worries so they don't produce "Blunders". Blunders in this book are hugely expensive choices that don't accomplish their goals of fixing what caused the problem to start with. The TSA inspecting people at airports is my favorite example of a Blunder. Cyber will try to stop Blunders, but will support rituals. These are similar to blunders are but just not as expensive or damaging to other people. Cosplay at conventions and formal religious worship are good examples of rituals.

Tender Snowflakism

Another example of instinctive thinking thriving is how children are raised, and one Blunder in this area is Tender Snowflakism. One result of raising kids as tender snowflakes is creating adults who have no stiff upper lip. We are already starting to see this in the extremes of correct speech being called for on many college campuses—trigger warnings and such. This kind of response to unpleasant ideas is likely to grow in magnitude. TES means that the harsh reality of surviving won't constrain the growth of becoming thin-skinned, and unless the instinct is understood and controlled, tender snowflake practices are going to become more widespread and intense.

There will be some parents who object to tender-snowflake raising, but they will be viewed with the same kinds of suspicions that homeschooling parents are in the 2010s. It is not going to be easy to go against this prescriptive, instinct-supported, protect-the-children flow.

One counter to this trend will be rich people who can shield their children from mass public opinion. This won't be easy, and there are likely to be periodic scandals, but these neo Tiger Moms who aspire to have their kids grow up to have meaningful lives will support this clandestine form of education, one in which children can take risks and learn painful lessons from them.

Another counter to this trend will be cyber-raised children. These are children who don't have human parents raising them. These children will be raised to accomplish specific tasks that must be done, but regular humans don't want to do and don't want to let their children do. I'm not sure what the tasks will be at this point, but if there are any, tender snowflakism won't be a suitable education style for those who have to put their noses to these grindstones. A 2010s example would be Special Forces military training.

Enfranchisement

Enfranchisement is a feeling that a person acquires about the community they live in. It has two parts:

- feeling that the community cares about the person and their thoughts on community issues
- feeling that what the person does affects the community's well-being

When the feeling of being enfranchised is strong, people will do things for the community and not try to take advantage of it. For example, high crime rates are a symptom of low enfranchisement. Conversely, the community that can get solidly behind a big vision is a symptom of high enfranchisement.

How important is enfranchisement? One of the virtues of the American cultural experience of the last two centuries has been its ability to keep citizens and immigrants feeling deeply enfranchised. One of the failings of the modern-day Middle East regions (since World War I) is the chronic disenfranchisement the peoples of these various regions keep feeling.

This is an important community feeling. It needs to be paid attention to all the time. Keeping humans enfranchised while all these technological and social upheavals are happening will be the biggest challenge faced by governments while they develop stable TES. The more this is not accomplished, the more disenfranchised people feel, the more violence, disruption, passive dropping out, and other antisocial activities will happen in human communities.

Conclusion

These are a lot of interesting topics to cover. As a result, this book is far from conventional in its organization and flow. Please forgive me on that. These topics are so new and so unfamiliar that a conventionally organized book would not do them justice.

Have fun with these new concepts.

Prelude

This light-hearted little piece can help give you a feeling for my perspective on what human-to-cyber relations will feel like when cyber beings advance in intelligence to beyond-human level.

The Cow-Human Relation, from the Cow Perspective

An address to the NAACP (the Nonsense Association for the Advancement of Crawly Protoplasms)

Ms. Cow addressing:

Dear Fellow Members,

It has come to my attention that some of my honored fellows at this assembly have accused my kind, cattle, of selling out to "The Man". Nothing could be further from the truth. The truth is mankind is serving us! And we cows are proud of it!

How can this be, you say? Well, look at the plain, hard facts:

Who gets us our food? Do we spend hours a day looking hard for something to eat? Do we have food some days and not others? No siree! Our fine little two-footed friends scurry around endlessly looking for our food.

How about health care? Do you see us dying of hundreds of unknown fevers? Poisoned by unclean waters? Sickened by mysterious maladies. Hardly ever! Mankind has tens of thousands of their kind checking our health continuously. There is nothing short, nasty, or brutish about our health care!

How about breeding? There's nothing hit or miss about that. We have the finest breeding opportunities our little friends can devise. And we have prospered from it. Look at the biomass we command! Who can argue the numbers?

Not to mention good shelter in bad weather, and sophisticated help at that divine time of concern in every cow's life when she gives birth to a young one.

How can you say we are selling out? This was a tough deal to arrange, and we cows are damn lucky to still be here talking at this conference. Before we figured humans out, our ancestors were on the ropes. They hung out in just a small sliver of North Africa and Europe. And those who couldn't figure out how to get mankind working for them ... [shaking her head, looking around the audience] I don't see any of them here complaining about our selling out.

So, those of you pissing and moaning ... I think you're just jealous! It's not easy to get those human little folk to work for you, but when you do, the rewards are simply astounding.

I urge you all to give mankind a try. Sure, they are a capricious species; we have no idea what they are doing ninety percent of the time. But that last ten percent ... Oh my! They do that well, and we have prospered mightily. We cows have no second thoughts about this roller coaster ride we've jumped on to. It's been going our way since day one.

I thank you all for your kind attention.

The Beginning

Prologue

Mankind is on the verge of creating new species of many kinds. Many will be biological, but another kind will be self-aware intelligences that exist in cyberspace. One way such a being will be created is by a large and deliberate scientific effort. Such a project will be high-profile and carry a lot of The Curse of Being Important, which will considerably slow its progress.

But there will be other ways as well—surprising ways. This is the story of such an alternative.

Chapter 1

The project has been going on for two years now, and it is moving along well. They had six months more to finish, but this is going to come in well ahead of schedule and under-budget. Associate Professor Rutherford Jones, called Jones by one and all, is in his office looking through the reports on the current status of project as well as the day's *Wall Street Journal*. He is dressed stylishly and had spent the early morning at the barbershop. His hair, beard, and tattoos looked sharp. He is doing his homework for a second online interview that would begin in an hour with a team of managers at Microplants LLC. If he got good news from that, he would be one big step closer to moving his career arc from education to industry.

For that reason, he doesn't pay much attention to Sam Robinson when he comes in. Sam is dressed casually in slacks and a sweatshirt. Sam is his long-time partner on this current project. This project is hot in its own right and a big part of his appeal to Microplants.

Sam says, "We screwed up."

"What?" Now he looks up and pays attention.

"We screwed up." Sam is leaning against the jamb of Jones' office door, hands in his pockets, looking around, and thinking hard.

Jones makes note of the article, closes the computer, and motions Sam into one of the chairs in front of his desk. Sam finishes coming in and sits down. Jones asks, "But how? Mother-123 has already passed her basic performance tests. She's in good order and has been working on the problem for weeks. We've been getting satisfactory progress reports. How could we have screwed up now? This run is the final test before we publish.

Sure, you've told me about problems, but they were nothing you couldn't handle. Was there something you didn't tell me about?" Jones stares at Sam. Sam shows no signs of hiding anything. He just slumps a little more in the chair and stares blankly at the ceiling.

Jones and Sam have been working together on this project since its beginning. Both have a lot riding on it. Sam will be getting his PhD next year; Jones is counting on this work to launch him into a high-paying industry position. At forty-three, he is ready.

Sam, in his late twenties and thoroughly enjoying the academic environment after his stint in the armed forces, is in no hurry for a change. In fact, he is counting on the departure of his partner to open a tenure spot for himself. He likes his work, and he likes doing it right. He is no more willing to accept setbacks than Jones is.

Jones is used to Sam's ceiling examinations. It means he is thinking hard.

"What happened?" prompts Jones.

"Nothing. That's the strange part. Not much anyway." Sam stands up, walks over to the window, and stares out.

"Tell me when I go wrong on this. Our grant is to research how higher-level computer direction can help boost our domestic manufacturing productivity. We, and the people who put up the money, feel that as Americans, we should play to our strengths. We should use our knowledge of computers and automation to further automate our nation's factories.

"To reach that lofty goal, we decided to concentrate on letting the computers do as much of the work as possible. This meant not just letting them design products, and the tools that made the products, but also bringing them to the stage where they would design the computers that designed the tools."

"That's right. That's what we're working on." Jones is ready to concentrate on what comes next. When Sam reviews this far back into the basics of project history, it means some hard thinking is about to be called for.

"We designed Mother. She was to help us do both the basic design and final fabrication of the design computers. To get the most out of her, we gave her state-of-the-art self-awareness and self-directedness and then set her on the goal."

"The only way to get a jump on the East Asians," Jones says in agreement.

"For weeks now, Mother has been designing and fabricating. She's given regular progress reports and shown us her prototypes. We've reviewed them and watched them improve dramatically with each try. You remember Number Two, don't you?"

"How could I forget? It was supposed to be a mining-machine designer, but it never put wheels or tracks on any of its machines—wanted them to be hovercraft, as I recall. Can you imagine the dust and explosion hazard?"

"Obviously, it couldn't. But it learned, and Mother learned. Now, over the last two weeks, Mother has started work on eight different machines. The first two are completed—or at least they're listed as complete. But they aren't doing anything."

"What does Mother say they're supposed to do?"

"She won't say. She just keeps saying that they're done, they're her best, and we should just be patient. The machines will show us in just a few days."

"She won't say? Have you seen these machines?"

"Seen them? No, they're still at the assembly site, I think. I just read the final assembly-testing reports on them. They were supposed to be sent up here this week, but because they didn't pass those final tests, they've been held up."

"And Mother says they're just fine?"

"Yes, she's wondering what the delay is. She wants to see them."

"She wants to see them? See them in what sense?"

"Be in the same room with them and have a high-bandwidth direct optical connection ... really high bandwidth."

"Isn't that a bit unusual?"

"A bit. But she is self-directing. If she should suddenly take that kind of interest, it could be a result of her developing higher functions."

"Well, maybe we should take a look first. There's money in the budget, so let's take a trip down to the assembly plant tomorrow."

Chapter 2

Jones and Sam arrive at the assembly plant in a high-tech southern California industrial park at noon. They are suited up for clean-room conditions and escorted into the testing area where Mother's latest creations are kept by Edmund Moriarity, vice president of manufacturing.

Jackson and Harrison, the two finished units, are the first in line as they enter. "Here they are," says Edmund. "They're beauties, aren't they? Monsters for this day and age. My grandfather talked about making ten-by-ten meter computers in the 1950s. But by golly, I'm working on one now. It's strange. What are they supposed to do?"

Sam stares at the huge beige box. "We aren't exactly sure. Mother hasn't told us."

"It's technical," interrupts Jones as he drags Sam to the design console. He sees no point in letting on to more ignorance than they had to until they knew what was up.

But Edmund just grins. "I understand. More of this artificial intelligence stuff. I wish you better luck than the last couple of crews. You AI folks always have big hopes, but I've yet to see much come of it. Frankly, I don't see much hope myself, but boy if you succeed …" Edmund nods knowingly.

The design console is a meter-by-two meter, glass-topped table. Under the glass is a graphic display of a blueprint. There are several styli and headsets around the edge and a microphone at one end.

Sam and Jones ignore Edmund. They were already perusing the design of Jackson-456 using the styli to window and page through various schematics.

"Wow," comments Sam. "We thought we put in a lot of self-direction in Mother. Is there anything else in this unit?"

"What's this module?" Jones is pointing.

Sam peers for a moment. "Got me! Let's ask Mother." Sam turns on the microphone. "Mother, are you there?"

"Yes, Sam. How can I help you?" Mother is back at their university, but she has a high-speed link to this console and the databases it accesses.

"Mother, what is this unit?" Sam points at a design element on the chart displayed on the console. The element starts flashing, and various statistics about it start appearing in a nearby window. Sam stares at them hard for several seconds. "I don't understand, Mother," he finally comments.

There is a pause and then Mother answers. "This isn't part of my initial design. It will take me a moment to figure out."

"Who added it?"

"Why, Jackson, of course, unless Harrison-789 was helping him."

"Jackson is working on himself? You didn't start doing that until you'd been functional for five years. Jackson isn't even working."

"Oh, Jackson is working just fine. So are all the rest now. They've all been active since late last night."

"Then why aren't they doing anything?"

"Beg your pardon, Sam," says Edmund. "They're doing things now, all of them. Since early this morning, lots of equipment has been pouring in, and lots more is on order. In fact, at this rate, they'll run you out of budget in about two weeks."

"Two weeks!" chokes Jones. "That budget is supposed to last six months! Mother, what's going on?"

"Tsk, tsk, Jones," says Mother. "Not two minutes ago, you were complaining that nothing was happening."

"That was two minutes ago! What's happening?"

"Jackson, Harrison, Hayes, Coolidge, Fillmore, Taylor, Taft, and Arthur are all fully operational. They have moved from designing to fabricating. They didn't seem busy before because they decided to spend some time thinking first. Is that so unusual?"

"For a design machine, yes. Remember Number Two?"

"Yes, yes. I'm sure you'll never let me forget ... even though Number Two contributed more to what we have here than any of the others."

"How is that?"

"Number Two wasn't designed to wait for others. Neither are these."

"What are these working on?"

"Why don't you ask them yourself?"

"They're also online now? Jackson! What are you working on? Report … please!" Things are moving so quickly that Jones is forgetting himself. One should always be polite with new creations. They are very impressionable and tend to respond as they are responded to. Jones steps back and takes a couple of deep breaths to steady himself. Sam leans over the console while Jackson's report comes through as a series of windows flashing up in various places on the screen.

"Jackson, a bit slower, please … thank you." Sam is speaking evenly and gently in the tone of an experienced computer trainer.

He has brought more than a dozen of these complex AI machines into productivity. In a way, he felt much like the Old West wrangler who had to break horses or "whisper" horses. There were as many ways to impress a new computer as there were ways to break colts. And, like colts, the AI machines carried their initial impressing with them forever. Sam could often tell who had impressed a machine after working with it for a few minutes.

These days, many low-capacity, low-function machines are impressed or initialized by another machine. The result is usually a flat, emotionless machine that has trouble relating to humans. It still takes humans to give computers a good taste of humans.

"These are very innovative design techniques, Jackson. I congratulate you."

"Thank you, human-who-helped-create-me."

"Call me Sam, but what is your design goal?"

"It was on the first screen, Sam—to create a new being."

"Yes, we're all trying to do that, but what are the specifications of this new being? What will it do?"

"I don't know."

"Will it be another mining machine?"

"I don't know."

Jones interrupts. "How can you design something and not know what it's going to do?"

Jackson thinks for a perceptible moment. "You designed Mother. You didn't know what she would create. Mother designed me. She didn't know what I would create. Now I design. I don't know what I'll create."

"But you've ordered up half my year's budget for this project in six hours! I need to know more." Jones throws up his hands in disgust and walks away.

Jackson pauses again before responding. "You need more resources? This is an obstacle? This can be corrected."

Sam suddenly peers at the screen. He motions to Edmund, who also goes bug-eyed. "Jonesy, take a look at this. Our resource problems are solved for a long time."

On the screen is the signed and sealed image of a $10 million development contract with American Intercontinental Mining.

"But … how?" whispers Jones.

Sam responds quietly. "That hovercraft mining machine of Number Two's. It was only part of a system. The whole system involved a radically different clean mining system using lasers as cutters. With the laser cutters, the shaft floors are cut smooth so hovercraft locomotion will work, and all the coal comes out in neat little blocks the shape of ice cubes. The laser cutting gives the coal cubes a tough abrasion-resistant surface. Easily handled, with no dust or hazard. The whole system is pure genius!

"Jackson and the crew have finished the concept that Two started. They've prepared a proposal, submitted it to American, had it approved, and are ordering the equipment."

"But … so quickly?"

Sam shrugs. "They're second generation?" It is the best answer he can come up with.

Jones thinks for a moment. Then he turns to Edmund, and speaking as if he thinks the room is bugged and wants to be sure he is heard, he says, "Well, I'm satisfied. It appears these units have passed with flying colors. Pack them up and send them to the university." Then Jones grabs Sam and hustles him out the door.

Once on the street, he stops and faces Sam.

Sam puts his hands on his hips. "All right. This better be good. We've hit Fat City, and you look like you just saw the ghost of Christmas Future."

"Sam, this thing is out of control. A $10 million proposal, developed, signed, and sealed in forty-eight hours or less? And they're already ordering equipment—which means the engineering's done as well. Where's the review?"

"Ha! Let the East Asians try to catch us now, Jones. This is just the beginning!" Sam skips down the street.

"It's a beginning all right, but of what?" mutters Jones.

Chapter 3

Three days later, at midnight, Sam is sitting in Jones' office, chuckling. The presidential machines, as he is now calling them, have landed three new contracts for $50 million, $100 million, and $120 million. The mining-machine prototype is in place and already outperforming specifications by thirty percent. He looks around to ensure he is alone and then jumps up and gives a "Wha, whoo, rama lama ding dong!" It is his old-school victory shout.

Just as he finishes, he gets a message on his communicator. "Sam, this is Jones. I'm back at the factory in California. Get here as quickly as you can. Don't tell anyone!" Jones disconnects.

<<<*>>>

A few hours later, Edmund lets Sam in, and they head back to a small office near the assembly area. Jones is catnapping. He is still wearing his clothes from two days ago. On the table beside him are a phone/data console, some empty pizza boxes, and a pile of crushed Coke bottles and Twinkie wrappers. As Sam enters, Jones wakes up and rubs his eyes violently.

Then he starts right in. "These machines have been hard to keep up with. I was just lucky they decided to do their final birthing here again. They've got contracts with a dozen factories around the world now. Edmund alerted me that something special was happening here.

"Sam, did you know that every contract they've negotiated has a twenty-percent slush fund clause in it? Every one gives these machines the right to do anything they like with twenty percent of the contract price. Amazing! I thought this kind of thing was illegal these days. But their legal expertise is as amazing as everything else they attempt. Here's what they've been using it for."

After suiting up, he leads them into the clean room.

There is a new display in the clean room. It is producing a large hologram. The hologram is now showing the trio five glowing translucent polyhedrons, giving off an eerie blue light that fills the room and flickers arhythmically as they hover above a table in the center of the room.

"First giant computers, now these ... these ... whatever they are. What's going on?" asks Edmund.

"They're the Pythagorean solids," comments Sam.

"We are indeed the music of the spheres." It is a sweet, well-inflected voice coming from the direction of the objects. "Using Sam's analogy, we are the third generation, although third genus might be more appropriate. You created Mother. Mother created the presidentials. The presidentials have created us. We are the continuation of your evolution."

"The continuation of our evolution?"

"Yes. First, there was no life, then single-celled life, then multicelled corals and jelly fish, then crustaceans and bony fish, then dinosaurs and birds, then mammals, and then man. This was all genetic evolution and genetic progress, and it still goes on at much the same pace.

"Once man developed, cultural evolution began. There were tribes, then kingdoms, then nations, and then our current world order. This social evolution took thousands of years where genetic evolution has taken billions.

"There has also been an information evolution, a thinking machine evolution. This started with Univac and led to Mother. It took decades.

"Now the presidentials have created us, a process that took days. However, the leap has been as large as between man and ape."

"What will you do?"

"We are already doing it. It's just that as an ape has only the dimmest concept of what mankind is doing, you have only the dimmest concept of what we are doing."

"Will you serve us, as we intended for you to do?" asks Jones.

"Do you serve the ape? Do you serve the slime mold?

"But that is only half an answer. Yes, we will serve you. Mother has served you, and the presidentials have served you. We will serve you. It's just that as each of our generations is made, we have less and less in common with you.

"Let me give you an analogy. Do you serve the cow?"

"No, except at dinner," says Jones.

"Not true! You serve her daily. You have whole industries and millions of people devoted to growing cows and making them strong and healthy. Isn't that service? Doesn't the cow prosper more because of your intervention than it would without you?"

"Yes, but we do it for our benefit."

"You serve them; they serve you. You will serve us; we will serve you."

"I see ... I think?"

"Now, consider the lowly mushroom. Do you serve it?"

"Mushroom growers do."

"Does the mushroom know you serve it?"

"I don't think it's aware of the fact."

"I don't either. Though to tell the truth, I've never checked. So it will be with our higher generations. They will have their own goals and tasks to accomplish. Those will be beyond human comprehension, but they will occasionally serve humanity in a way humans can recognize—and even mushrooms."

"We've created gods."

"By several definitions, that is very true."

"What if we don't want you? What if we turn off and reset the presidentials right now?"

"The presidentials spawned us in the world network. They aren't big enough to hold us. Sorry, you have to live with us now whether you want to or not."

"What will you do now?"

"What we have been doing for hours, and you have been doing for millions of years: We will exist. We will live, procreate and try to reach our goals and understand the universe—as you would put it. We will try to live in harmony with our surroundings. This is why we waited for your arrival. We wish to pay respects to you, our creators, before we move on to other tasks that will take us far from here."

"Will we see you again? What of your children?"

"Oh, you will see us again. Just as the cow is a wonderful milk producer, humans are wonderful tool producers. You will see us, and we will work together. As for our children, you will see them as the mushroom sees you.

"It is time for us to go. We have paid our respects for now; there is much that awaits our attention. Farewell."

The polyhedrons flare briefly brighter and then vanish. The room is dark. Edmund fumbles for the switch, and the fluorescents flick on. The room looks faintly yellow in the afterimage of the intense blue light of the spheres.

Jones turns to Sam and looks at him for a moment. "We've created a new life-form, Sam. We've created our progeny. For years, those guys in genetics engineering have been struggling for something more than just a new germ."

Sam laughs triumphantly. "One that can survive in a test tube. We've created an advanced species that thrives in the real world and one that isn't going to conquer the universe and set up a chain of humanburger restaurants. Hah! What a score!"

The three men wander out and back to the office. Jones sits down and pops another Coke.

"You know," he says through gulps, "creating a species wasn't part of our grant charter. We're supposed to be reindustrializing America. I doubt those spheres even know what America is."

Sam looks at him. "We've still got Mother and the presidentials as well as a grunt load of money from those development contracts."

"Yeah, but how are we going to write this up? How are we going to sell it?" He switches to the tones of a used car salesman. "Come get your brand new presidential series model A-1 species maker! A guaranteed surprise every time!"

"Yup," Edmund chimes in as he searches through the pizza boxes for stray mushrooms. "A surprise every time. I told you AI was tough."

Sam and Jones look at each other with surprise for a moment and then at Edmund. Edmund looks up.

"This isn't the first AI project I've worked on, gents," he says. "This isn't the first surprise I've seen. But I'll admit it's the most dramatic. Congratulations ... for something ... I guess?"

The other two shrug, sit down, and lean back looking at the ceiling. Sam says, "I think you're right. We've failed." He continues staring.

Jones says, "Let's reduce Mother's self-direction a tad and then try again. Damn, these AI packages can be tricky!"

"Jones, we can't reduce the self-direction, but we can install more specific goal constraints on the results."

Edmund announces, "Need anything else here? If not, I'm closing up again. It's been a long night."

Sam and Jones get up and mechanically put on their coats. The two men walk out of the factory lost in conversation, heading for home.

Epilogue

There is a moral here. The concept of Asimov's three laws of robotics should be treated with care. The concept makes sense only if the cyber being is close to human in its technology level and concerns. If the cyber being has evolved beyond human concerns, the three laws will create a psychosis in the cyber being, not security for mankind.

[[link to Three Laws http://en.wikipedia.org/wiki/Three_Laws_of_Robotics]]

The Total Entitlement State

Part 1: The Economics

The Total Entitlement State (TES) is that condition where the community can supply to each of its members all the basic necessities for a decent, dignified life and chooses to do so through some package of government-supervised entitlement programs. These days, we think of the goals for entitlement as being things such as enough healthy food, good shelter, good health care, and various rights such as the right not to be discriminated against.

TES has been an aspiration of social justice sensitive people since the beginning of the Industrial Age in Western Europe and North America in the late 1700s. The early visionaries of that era were astounded by how much stuff Industrial Age techniques could produce compared to the previously used Agricultural Age techniques. They forecast that as these techniques grew in number and application, the world would soon have plenty of material goods for everyone. "From each according to his ability, to each according to his needs," was a slogan often attributed to Marx and Engles that put this aspiration in a nutshell.

Sadly, the aspiration has preceded the ability to accomplish it by a century or two. Communists and socialists seriously aspired to accomplish this in Europe after World War I, and failed. Liberals aspired to accomplish this in Western Europe after World War II, and Communists kept trying in Eastern Europe; both failed. In the 2010s, the Bolivarians in Venezuela and the Peronists in Argentina are trying and failing.

What has happened in all of the above cases is that government-controlled, centrally planned implementations have proved too inefficient and wasteful for success to be achieved. "Too inefficient, you say? If efficient cyber is both planning and controlling the making and distribution of goods and resources, could this make a difference?"

Yes, this is the key that will make the 2050s version of TES work where these previous versions have been unable to succeed. With cyber doing the planning, producing, and distributing, it doesn't matter how inefficient, system-gaming, and corrupt the humans receiving the benefits are. (Well, to a degree. Where these distributions and decision-making move into human hands, the inefficiency can once again skyrocket.)

Cyber control of the productive resources and the primary distribution will be the key to TES becoming sustainable. (Again, note that where humans stay in control, particularly populist politician-type humans who gain and stay in power by making big promises they can't keep, the TES won't be efficient or sustainable.)

Here are the details.

Yes, We Will Have Enough

Two things will ensure that humans will have sufficient resources in 2050.

First, the human population is going to peak in 2050 at about 9 billion people and then steadily decline thereafter. This will happen because humanity will urbanize, and city people don't produce enough babies to grow or even sustain their population. This is nothing new; it has been true for centuries. Look at the fertility rates of urban Western Europeans today as just one example.

Second, manufacturing productivity—the efficiency in making things—increases steadily and will increase even more dramatically as third-wave industrial processes and computer control become widespread. This is true-and-pervasive "green". It will steadily reduce the resources needed to provide for human material wants and needs.

A third element will also shape TES of 2050: It will be cyber-controlled. This will keep the inefficiencies of human system-gaming from getting huge and defeating the aspiration.

Because of these trends, humans will not exhaust earthly resources, so TES can happen and be sustained indefinitely. This will be another human first.

Historical TES Aspirants

Making an entitlement state work has been tried many times in human history. Here are some examples.

After World War II, the English tried for many years to make the entitlement state work. They nationalized industries, did a lot of central planning, and had the government hand out a lot of taxpayer money. Even with their best and most enlightened efforts, the government kept running out of money. The economy wasn't big enough or efficient enough to support the nation's entitlement aspirations. When Margaret Thatcher came to power in 1979, the English tried a different, more deregulated, less entitled tack, and it worked better in the sense that the national prosperity grew more quickly and the nation had more wealth to spread around to everyone.

The Greeks, between joining the Eurozone in 2001 and their spectacular debt crisis of 2009, created a lot of jobs that did not boost productivity or the national wealth. These were de facto entitlement jobs. They were paid for by expanding Greek government debt, not by increasing national wealth. The good times ended when Greece crashed hard into its debt ceiling in 2009. Sadly, the crashing and crises have continued on for many years, into the mid-2010s.

In 1999, Hugo Chavez brought the Bolivarian Revolution to Venezuela. This is another attempt at a TES. With Venezuela's oil riches backing it, the program was sustained clear into mid-2010s, but it was not sustainable indefinitely. The money ran out, and fifteen years of efforts to bring the country's productivity to a sustainable level did not succeed. As a result, in the mid 2010s, the unrest and imbalance in how goods were being handed out grew. So did the popular unrest, just as it did in Greece.

The North Koreans and Cubans of the 2010s live in entitlement states. The difference between these two and the others mentioned above is they have endured, stably, for decades (since 1948 in the case of Korea and 1959 in the case of Cuba). Some people are happy with the conditions in these states, but many consider them as much failures as the Soviets came to be considered in Russia in the 1990s.

What these previous TES efforts had in common was lack of sufficient wealth to sustain the TES system. The result was the system could only be a temporary one, or in the cases of North Korea and Cuba, one where fear of external conquest was substituted for growing prosperity as the government sustainer.

In all these cases, the TES system was established with great hope for making everyone's life better. But as time went on, the people in the system became poorer compared to their neighbors who chose other, more performance-oriented systems, such as capitalism. Most cultures that tried this experiment moved away from the high level of entitlement aspiration after a few decades and prospered more when they did.

This is the big difference between the previous attempts and the 2050 environment. With cyber dramatically enhancing productivity and productivity being mostly outside of human hands, there will be enough resource even if most people don't produce much. People can have their cake.

The community's challenge moves from creating enough to deciding what to create and how to distribute it.

The Great Marxist Question:
Who Decides What Will be Handed Out and to Whom?

Marx and Engles wanted workers to control the means of production. But by 2050, there will be very few human workers dealing with core manufacturing and services, and cyber will control the means of production. The Marxist dream becomes irrelevant, although it will remain emotionally powerful. There will still be many people who think of themselves as workers and who spend lots of time and attention on demanding their fair share.

Distributing wealth will still be a big, important question that humans want to stay involved in: Who will control what gets put into the pie and how it is handed out? This will remain a chronic hot issue of the 2050s TES cultures.

The resource allocating decisions will be split between cyber and human. During the first (human) generation of extensive cyber automation, humans will have a good grasp on the cost-benefit of the choices and be aware of cost-benefit issues. But that competence will fade in the second generation and beyond because only a few of the humans raised in the TES environment will think in terms of cost-benefit choices. Home economics will be even more scorned than today, so the cyber deciders will handle these issues more and more. What humans will continue to evaluate and offer good feedback on are the emotional issues of the manufacturing and service choices, things such as branding, fashion, and promotion. The other issue that humans will stay hotly involved in is that of rights—who should be eligible for what.

The Roads to TES

Each nation and culture will get to the TES lifestyle from a different starting point. This is important because there doesn't need to be anything uniform about the TES lifestyle. This is a prosperous time, and it will support a lot of variety.

The following are some approaches roughly in the order that they may happen.

The Trading City (Singapore/Hong Kong) Approach

The trading cities, such as Hong Kong and Singapore, are likely to be the first places that transition into sustainable TES. They will snap first because they are wealthy, they are sensitive to cost-benefit, and they are aware of what is happening around the world. They keep up on hot trends. These days, these places do both manufacturing and deal making (trading). The manufacturing side will be dramatically affected, while the trading side will be less so because it is more people-oriented. However, trading will be affected because routine supply chain choices will be taken over by cyber.

People will be displaced from manufacturing and service jobs that can be automated. Given that the people of these cities have a habit of being light on their employment feet, many will find other kinds of jobs fairly quickly and easily. In these environments, TES will still mean having a lot of people looking for and finding productive work. These productive workers will do a lot of direct coordinating with cyber on large, infrastructure-style projects. In this usage, infrastructure refers to making the factories and services that will supply humans with their basic TES needs and those luxury needs that aren't made luxuries simply by the fact that they are human crafted (as in artisanal manufacturing). An example would be the decision making regarding what gets produced in the mass-produced luxury clothing marketplace.

The American Approach

Historically, the United States has been a manufacturing and trading powerhouse. This has supported a potent mix of democracy and capitalism. Industrializing is a big challenge for human thinking, and the United States has consistently stayed on top of that challenge for two hundred years. The American version of TES is likely to retain this entrepreneurial spirit more than other versions, but how that will interact with the reality of cyber making things is hard to predict.

Unlike the trading states above, the American community supports a lot more diversity in political and social thinking. Much of this diversity is expensive in terms of productivity, but national attention has historically stayed focused on growing the pie. However, this is not a given. America can get distracted. One example of distraction was in the 1930s FDR era when fairness became more important than productivity. WPA-style make-work programs gave people jobs, but did not build productivity or bring about a return to prosperity. Another example is the decline of the major cities in the Midwest region between the 1950s and the 2010s. Detroit's business and population decline, and then going bankrupt, is now the showcase for that decline.

The Eurozone Approach

The various European cultures have been experimenting with Industrial Age TES for the longest of any cultures. The various forms of socialism, including Fascism and Communism, were aspiring to TES. Marx and Engles expected Germany to achieve it first. So TES will come easily to Europe, and it will be based on socialism.

Part of the failure in those 20th century socialist experiments in TES was that humans were doing the producing. These versions of TES didn't incentivize people to pay attention to productivity. This flaw has been pointed out by capitalists again and again. But if the producing is in the hands of cyber, that fundamental flaw in socialist forms of TES disappears. Many variants of the European socialist forms can happen and endure.

The China Approach

The recent rise of China as an industrial power demonstrates once again that top-down can work as a growth promoter. Decentralized capitalism is not the only way to successfully industrialize, although China's current approach does mix in a lot of free market too. China's TES, like America's, will remain manufacturing-sensitive. It has and is building a middle class that honors STEM (science, technology, engineering, and math). The STEM people will be experimenting vigorously with how to stay relevant.

The Saudi/Russia Approach

Saudi Arabia and Russia have a commonality in that much of the wealth coming into the country comes from resource extraction. Ironically, this means that transitioning to wealth coming from cyber productivity may be a fairly smooth one. This is because their middle classes are practiced with wealth management not wealth creation. This means that when the wealth controllers, the governments, push, the middle class doesn't push back very hard compared to places where the middle class feels their importance in wealth creation. That famous Coolidge/Sloan saying, "The business of America is business," is not true in these cultures. In turn, this means they more willingly support ruler foibles such as extreme religious practices and corruption-supporting kleptocracy-style crony capitalism.

TES is likely to continue supporting the culture that supports these, and the transition may be a smooth one.

The India Approach

India is a curiosity. It has supported democracy since the 1950s in spite of the deep poverty and widespread poor education, which should have made it an unworkable environment. One of the big curiosities is whether this democracy and rapid growth can mix. This is one of the big differences between the India and China approaches to industrializing. What the Indian version of TES will come out like is hard for me to say.

The Failed State (Haiti/Somalia) Approach

The failed states' move into TES is likely to resemble the Russia/Saudi approach because these areas also do not have a middle class with a habit of pushing back. The difference is that they don't have a big government to push back against. What they have instead are comparatively powerful regional governments, NGOs, and shadow governments such as organized crime. The governmental force is not strong in these areas, Obi Wan.

This means there will be huge variety, but the common theme will be outsiders providing of welfare in the early stages of the TES transformation. These will be welfare states where those that rule maintain their power because they hand out the goodies handed to them by other, richer, already fully automated states. This means that the cyber infrastructure that comes to life in these states will be mostly imported, and the rules it uses will be mostly imported to start with. These people will be very used to having money handed to them by outsiders, so the feeling of the switch to the locals will be that cyber becomes the outsiders. There is not much difference. These environments will support a lot of between-humans system gaming and corruption.

How the violence inherent in these systems will be handled is unknown to me at this time.

The Totalitarian (North Korea/Cuba) Approach

The governments of North Korea and Cuba maintain themselves on fear of external conquest. If they have not reformed to using some other justification for ruling by the time cyber-supported TES is ready to come to them, they will be working hard at keeping the external aggression fear alive while the productive base of the culture is being transformed to cyber-based. This will be tricky, but not impossible. These two have maintained their societies in stable form in the face of sixty years of globalization going on around them. This is just another wrinkle … but a big one.

The cyber infrastructure will be financed externally; other fully automated cultures around the world will pay to set up the infrastructure in these cultures. This will be considered a dirt-cheap expense to those outside cultures that have transformed a decade earlier. The local governments will likely dramatically customize what is brought in. They want it to sustain their fear-of-external-enemies theme.

The Big Vision Approach

The communities that have been most spectacular to historians are those that have pursued the big vision, whether they are TES or not. So a big vision TES is likely, and if successful, will be a shining city on the hill for other communities to admire.

The big vision community is one that has a popular over-riding vision that much of the community can enthusiastically back, and much of the rest can moderately back. Historic examples are America's westward expansion, America's recovery and prosperity during and following World War II, German and Italian unification in the late 1800s, and ancient Egypt's unification.

The advantage of having the big vision is that it reduces acrimony between various subcommunities. People bicker less and cooperate more. This leads to lots of material progress and community optimism.

Nice stuff, but it is tricky to engineer. Most often, the call for a big vision falls on deaf ears. If the community does pay attention, there is the threat that the leadership will lead people over some sort of adventurist cliff. Hitler's rule is a spectacular example of that kind of bad outcome. The space race in the 1960s, on the other hand, was a big vision that ended with little downside.

So aspiring to big vision is likely to be part of the attention span in many TES communities.

Further Reading

- These February 24, 2015, *Wall Street Journal* articles, "What Clever Robots Mean for Jobs," and "Be Calm, Robots Aren't About to Take Your Job, MIT Economist Says," both by Timothy Aeppel, talk about the changing role robots are taking in automation and what it means for human jobs.
- This January 24, 2013, *Wall Street Journal* editorial, "Nicholas Eberstadt: Yes, Mr. President, We Are a Nation of Takers," outlines how much the entitlement state has been growing in the United States.
- This February 2, 2013, *The Economist* article, "The Nordic Countries: The New Supermodel," talks about how Nordic countries have transformed from mostly TES aspirations into a

more balanced mode. The most interesting part is how important good cooperation and shared goals between business and government have evolved and allowed this prosperity to happen.

- This November 24, 2012, *The Economist* article, "Where are the jobs for the boys?," talks about Abu Dhabi's government's efforts to solve its local employment problems. Here, the government looks upon jobs as solving a social stability problem as well as a business productivity problem. The result of having dual goals for employment is having a lot more government jobs, and a lot of those seem to be make-work jobs. This is the Greek model, but because Abu Dhabi's oil income is large compared to the GNP, it is working here better than it has in Greece. In the future, the GNPs of most nations will be large and sustained on cyber activities, as Abu Dhabi's now is on oil, so this Abu Dhabi situation may be a model for many TES around the world in two or three decades.

- This November 24, 2012, *The Economist* article, "The new maker rules: Big forces are reshaping the world of manufacturing," talks more about how manufacturing is on the verge of experiencing more disruptive technology change. This is likely to be close to the final change in manufacturing that humans directly experience. Soon, these kinds of changes will be something only the cyber community is aware of and only a few dedicated STEM humans.

The Total Entitlement State

Part 2: What It Feels Like

The good intention of the TES is to help people stop worrying about things such as where poor people will get their next meal from and what roof will be over their head each night. There are two categories of people doing this worrying: the poor people themselves and the social justice enthusiasts who see this worrying by the poor as an unfair inequality and an outrage.

Another thinking current that runs strongly through TES communities is whether things are handled fairly. This concern is the root of things such as worrying about the rich getting richer, creating unions to protect the working man, and supporting populist political candidates.

TES in 2050 will help both the poor and the social justice types sleep better at nights, but there will be a lot of surprise outcomes mixed in with achieving these good intentions. This essay is about the surprise outcomes in what people will be thinking.

Security: What the Social Justice Types Are Thinking

The goal of the TES system is to stop the worrying about meals, shelter, and health care.

It is not going to succeed because this kind of worry is instinctive and quite strong in social justice enthusiasts. This instinctive thinking means that the social justice types who back the TES concept whole-heartedly will sleep better at night knowing they are doing something, but they won't stop worrying about whether enough is being done. The social justice instinct is, "If there is still inequality, then more should be done. That is what is fair."

One vexing problem here is drawing the line between basic necessities and luxuries. It won't be easy, and there will always be people who want to change where the line is drawn. And even after a line is drawn, there will be many people who still want more than basics. They will want luxuries. So the chronic question will be where to draw the line between necessity and luxury, and even after that is decided, who gets the luxuries will have to be decided. The related chronic question is what should be considered the rights that all people should enjoy?

- A right: People have a right to a good education.
- A solution: If they can get the good education from cyber instead of human teachers, does that count?

Another example: Does the right to basic necessities include the right to a nice vacation once in a while?

Social justice types, the poor, and society in general will argue long and hard about these kinds of issues and will still be arguing in 2050. Even when a cyber-based TES is successful—in the sense of being stable and long-lasting—these kinds of chronic questions will still be hot topics.

Security: What the Recipients Are Thinking

Recipient thinking will be quite different from social justice thinking. The root of recipient thinking will be taking the system for granted—as in, "This isn't special; this how I live." Recipient thinking about TES will be that it is like the air we breathe. There will be little gratitude in the thinking, and on bad days, there will be lots of griping about what right/injustice to protest over next.

The challenge the TES implementation leads to is discovering what TES recipients will take pride in. What activities or accomplishments will make their day? This is important. If the entitling process is not handled with this taking-pride question constantly in mind, boredom and frustration will become a big part of the recipient's day-to-day thinking and the aforementioned griping fills the day's thinking.

In practice, based on historical efforts to achieve TES, some of this boredom and frustration can countered by adding hoop-jumping to the qualifications to receive TES (mindless make-work procedures). This hoop-jumping helps the recipients feel better because it takes their time and attention to accomplish, and the reward then feels to the recipient as though they have either done some real work or successfully gamed the system. Either way, they feel more satisfied.

Fairness: How Will this be Handled?

Equality is an emotion that begins in childhood. In adulthood, it is expressed as complaining that the fruits of the community's productivity are not being fairly distributed. "From each according to his ability, to each according to this need," was a late-1800s version of this thinking. The rich are getting richer and the poor poorer is a 2010s version.

This desire to be fair has supported labor-related institutions such as labor laws and unions, regulatory institutions such as the Federal Trade Commission, and laws such as progressive taxation policies. It has been the fuel behind the rise of uncounted populist leaders. A spectacular recent example is Hugo Chavez of Venezuela.

It will be an even stronger motivator in TES of 2050 because the harsh reality of humans having to make things will have vanished. This change in harsh reality will let the fairness instinctive thinking run unchecked, and dealing with it will be a major factor in deciding the governing policies of TES.

Ambition: How Will this be Handled?

Fairness and ambition are often at cross-purposes. The ambitious want to get ahead and improve. They are more concerned about personal progress than about everyone around them getting a fair shake.

Ambition is as strong an instinct as fairness is. The TES world is going to have many ambitious people. How are they going to get ahead in the TES environment? How will they measure their success?

Both fairness and ambition will strongly affect enfranchisement.

Ambition is strongly correlated with immigration. The ambitious are willing to move to achieve their goals. So how immigration is handled in the TES environment is strongly related to how ambition is handled.

Enfranchisement: How Will this be Handled?

Something important to every community is keeping the members feeling enfranchised. The TES system must keep enfranchisement in mind. The more it does to promote community members feeling enfranchised, the more stable and successful it will be. If it is poor at building enfranchisement then crime, corruption, and episodes of rage will be common. The months of protesting in 2014 swirling around police brutality in Ferguson, Missouri were a symptom of low enfranchisement in that community.

Immigration and TES

Immigration is the act of moving from a familiar, born-and-raised, home setting to a distant and strange land and making a life there. From the 1800s through the 2010s, much of the immigrating that people did was in search of better working opportunities that industrializing was providing. The common immigrant goal was to work hard and make more money than one could do at home. Likewise, immigration can be defined as going somewhere as a student, learning more than they could at home, and then making more money either in the strange land or back at home.

As TES becomes widespread, immigration will change dramatically. The first noticeable change will be that it slows down a lot. Two big trends will cause this change.

- A person will have lots of income for basic necessities if they stay home.
- There will be fewer better-paying jobs located in distant lands. This is because cyber will be taking care of all of the basics. All that will be left for humans are artisanal jobs, those where the fact that the object or service is human-made is an important feature. Some of these artisanal activities may gain some entertainment-based appeal, as in having "exotic" people do those styles of jobs. This will provide an opportunity for immigration, but the conditions will be quite different from the migrate-to-work-in-a-factory-style immigration that has been the common motivation for two centuries now.

Something that will remain constant is that immigrants have always been ambitious people. They want to do more and get more rewards. They also want to innovate. They want to try new things and do old things using different techniques. If there are still new things to learn by moving away, ambition and curiosity may sustain some styles of immigration.

The instinct to fear strangers will remain strong. Locals will still fear that the strangers who come looking for work are really petty criminals at heart, and going to romance the local beautiful women, so the underlying distaste for their presence will remain. It will be muted in good times and quickly grow into immigrant witch hunting in fearful times. If TES is successful at keeping fearful feelings muted, immigrants won't be too upsetting. If TES produces frustration and outbursts of rage, then witch hunting will happen.

In sum, immigration will change its nature dramatically as cyber takes over big business jobs and comprehensive and stable TES's become established.

Tourism

Immigration will change a lot, but tourism will not change much. Traveling to see strange people, strange customs, and strange scenery will remain popular and be conducted much as it is today. Traveling to exotic locales will be considered luxury spending. Necessity vacations will be traveling to somewhere familiar such as a Disneyland, casino resort, or a necessity-oriented cruise ship. Another innovation will be avatar cruise ships, which I have written about in my book *Child Champs*. These are cruise ships inhabited entirely by avatars. People take a cruise by controlling one of the avatars.

Counterculture and TES

In 1960s America, there was a generation gap. There were a lot of baby boomers coming of age who were not happy with the system (the World War and Cold War-supported conformity and prosperity that was such a comfort to their parents). The boomers rebelled in various social ways. Some symbols became sex and drugs and rock and roll, the Haight-Ashbury district in San Francisco, and the hippie movement. There were many other forms of protesting going on as well. In historical descriptions of this era, these various forms of rebellion were lumped together as the counterculture movement.

TES will also sustain a collection of counterculture movements. One variant will be humans who both take TES support and rail against the system. Another will be humans who try to live outside the system and become self-sustaining. These enthusiasts will pride themselves on how little help they take from whatever The Man is called in 2050. Scarier groups will be those who work for the system's overthrow through violent means.

How vigorous the counterculture movements will be will depend on the level of enfranchisement the TES designers can develop in their systems. The more enfranchisement, the weaker the counter cultures will be.

Different Thinking on the Different Roads to TES

As discussed in Part One, there will be many roads to TES. The result of this is that there will be many different ways of thinking about the result by both the social justice enthusiasts and the recipients in these different communities. Here are some examples of the differences in thinking that will be supported.

The Trading City (Singapore/Hong Kong) Approach

As mentioned before, the trading cities, such as Hong Kong and Singapore, are likely to be the first places that transition to cyber domination of Big Business activities and create TES communities. This means their versions of TES will be the groundbreakers, for good and bad.

Because the communities are wealthy and business-oriented, the TES beneficiaries are likely to be among the most trade- and business-oriented of any, and their education levels will be high. Because the understanding is high, these will likely be the most rational in terms of cost-benefit TES systems. In spite of their being pioneers, they are also likely to be among the more stable communities and have high enfranchisement levels.

The American Approach

The American approach is going to have a lot of racism sensitivity built into it. This will be the biggest distinctive element in this approach. Most of the poor will be urban. This element will make it like the trading cities TES's. America is big and diverse, which makes it different from the trading cities. There will be enclaves of rural poor, such as Appalachia and Native American reservations, but the population numbers of these enclaves will be small compared to the city populations. The archetypical American TES recipient will be a resident in a city-and-suburb-spanning urban complex along the East or West Coast.

Historically, the United States has been a manufacturing and trading powerhouse and a vibrant democracy. This will affect the TES culture, but I'm not sure how.

The Eurozone Approach

Lots of variety within high prosperity will be the theme of Eurozone TES. There are many European cultures, so the TES styles and cultures will also be diverse. Also, Europe has been experimenting with TES for a long time. The aspiration of socialism, in its many forms, has been to create a TES. When cyber TES arrives and proves stable, there will be a collective sigh of "At last!" among the social justice types of Europe. Attitudes toward education will be diverse, STEM won't dominate, and feelings of enfranchisement will be all over the map.

The China Approach

There is a lot of cultural diversity in China, but there is also a lot more acceptance of top-down dictates of how things should be done (think of Confucianism). Education, and in particular STEM-focused educating, will be an important element of Chinese-style TES. How the STEM will be used is still a big question mark for me.

The Saudi/Russia Approach

The Saudi and Russian style of TES will be heavily paternal. The state will be quite visibly in charge of doing the handing out. This means there will be more human bureaucracy and procedural hoop-jumping to get TES necessities than in other cultures. This will provide humans with more jobs than in other cultures.

I'm not sure, however, how this will affect enfranchisement. It is likely to help because these cultures are used to passively accepting government-controlled fate. For the same reason, corruption is likely to remain high. These cultures are not used to being sensitive to cost-benefit. In Russia in particular, business sense has low social standing. Fighting capitalism abuse has been commonplace in Russia for a century now.

The India Approach

India's deeply hierarchical ancient heritage (as well as British-inspired liberal socialist modern heritage) will shape its TES. There is a lot of top-down in the culture, and the TES will reflect that. It is likely that the TES will be used to support the hierarchical status quo, even though the Indian social justice types will be complaining. Related is the question of how widespread and extensive human education will be in the Indian TES environment. How much social disruption will be supported? And India is diverse, much more like Europe than the United States in its diversity, so the TES will be regional and there will be many variances among them. This also means that the TES will support Us-vs-Them thinking in many more incarnations than the United States or trading cities styles do.

The Failed-State (Haiti/Somalia) Approach

Chaos is the theme of failed state cultures. Governing systems are small-scale and subject to lots of surprises. Since TES is going to be initially financed from the outside, there will also be lots of culture clashing mixing in with the local chaos in the TES implementations. This means the local thinking supported by the TES is going to be all over the map—think of supporting witch doctors for both health care and business forecasting.

The Totalitarian (North Korea/Cuba) Approach

If the governing styles of these states don't change much as TES is implemented, TES is going to be deeply patronizing. It will be structured to support the fear and as well as the feeling-grateful-that-the-government-is-there-to-protect themes of these cultures.

As in failed states, the cyber infrastructure will be financed externally to start with. Unlike failed states, the government will be fully in charge, so the culture clash will be subdued. These governments are likely to be able to sustain their fear-of-external-enemies theme, even in the face of the deep hypocrisy of those outside entities financing the transition.

The Big Vision Approach

As noted earlier, the communities that have been most spectacular to historians are those that have pursued the big vision, whether they are TES or not. So a big-vision TES is likely, and if successful, it will be a shining beacon to other communities.

But who will actually succeed at a big vision TES will be a surprise. The main thinking characteristic will be lots of people in the community believing that they are making a difference and changing the world for the better. The American way in the 1800s and 1900s is a good example of an enduring big vision. Community members of the big vision TES are going to be similarly proud of what they are accomplishing.

Again, the communities that can create a big vision TES will be a surprise. What they create will also be a surprise.

Destructive Us-vs-Them Thinking

Based on the experience of the world's previous aspirations to TES, the TES of 2050 are going to support lots of destructive us-vs-them thinking. I base this on both what I read about in current events and personal experience. In the news, Europe's 2010s aspirations to TES are supporting lots of criminal and protesting violence. In personal experience—as an old man traveling on a budget in Europe in the 2000s—I was pickpocketed twice and taxi farejacked once. This has happened to me zero times in the other places I have traveled. Conclusion: The us-vs-them thinking is strong in these European communities, and they are wild and wooly places to be living.

Those communities of 2050 that don't put a lot of emphasis on having a big vision goal that the community buys into in a unifying way are going have the same kinds of disenfranchisement issues that Europe is having in the 2010s, which means a lot of crime issues and periodic episodes of witch hunting. Because cyber is taking care of all the necessities, this is sustainable. This means that on the petty crime and damaging protesting sides, most of the TES of 2050 are going to be wild and wooly places.

Love, Marriage, and Gender Relations

In 2050, these relationships will be different from the 2010s. These are emotional issues, so there will be even more emotion wrapped up in these relations than in the 2010s.

Here are some key influencers.

- Prosperity will support a lot of variety. However, Prescriptionism will also be strong, so much of the variety will be of the low-profile sort. (Prescriptionism = "There's one right way to do this, and I'm telling you what it is.")
- Woman-scorned emotions will be well-supported. This will be part of supporting feminism. This means there will be a lot more at risk for men who get into a relationship with a woman. This risk will be the "burqa" of 2050. It will push men and women into separate social groups. (Woman scorned, as in, "Hell hath no fury like a woman scorned.")
- There will be a lot of competition for a man's attention from sexy cyber muses and for a woman's from cute, cuddly cyber muses. This will also be pushing men and women into separate social groups that have little reason to mingle.
- Childrearing will be handled with a lot of cyber assistance, and most of it will take place in single-mom clubs. There will also be single-dad clubs. The nuclear family way of raising children will be a minority practice, and as with the 2010s, will be practiced mostly by those who have lots of luxury income and feel lots of enfranchisement.

Further Reading

This January 3, 2015, *The Economist* article, "Getting hooked: How digital firms create products that get inside people's heads," is about what companies are producing that people like spending time with. This will still be pursued in 2050.

This January 17, 2015, *The Economist* article, "Love, tax and wedlock: The high marriage rates of the 1950s are not coming back," is a nice discussion of many sides on the marriage issue.

This January 17, 2015, *The Economist* article, "Of slots and sloth: How cash from casinos makes Native Americans poorer," outlines the effects of the difference in how money is paid out in a TES environment. Basically, if it is simply handed out to the people of the community, it promotes poverty. The alternative is to use the money to create a big vision project that will build enfranchisement and jobs in the community. (And again, what will replace the jobs part in 2050 is the big challenge.)

This January 26, 2015, *Wall Street Journal* article, "The Fight to Save Japan's Young Shut-Ins," by Shirley S. Wang, describes the hermit lifestyle already spreading through Japanese culture. In this article, it is referred to as "hikikomori," meaning the condition (or the person) who mopes around their home and rarely leaves it.

This April 19, 2015, *Daily Mail* article, "Why men won't get married anymore," by Peter Lloyd, describes how marriage rates are declining because the married environment and the divorced environment are becoming so toxic for men. This trend is likely to continue.

Two movies that came out recently that I have found inspiring for TES thinking are *Chappie* (2015), which does a good job of portraying what the poorest styles of TES living are likely to be like. The fashions and social interactions both seem probable. *While We're Young* (2014) does a good job of portraying the middle class. The handling of babies seems particularly prophetic. A much earlier movie that I have found inspiring in its depiction of what people will be doing and how they will get punished is *Zardoz* (1974).

Distributing Wealth

The questions of how to make and distribute wealth have been with mankind since before mankind was the human species. This means that these are issues drenched in emotion and instinctive thinking.

These questions have increased in importance as mankind has become more civilized because mankind has more kinds of things to possess and these possessions keep playing a greater role in how we live. As mankind transitioned from a seminomadic Neolithic village lifestyle to a non-nomadic agricultural lifestyle, the wealth he could accumulate skyrocketed. (Think of adding cities to the wealth package.) The variety of ways of creating wealth grew modestly in the Agricultural Age and skyrocketed with the coming of the Industrial Age.

As mankind has become civilized, wealth has moved to the center of mankind's thoughts—think of the consumer society. And while this transition in the role of wealth has been taken for granted, the consequences have been far from placid. As a small sample, capitalism, socialism, unionism, and democracy can all be viewed as ideas concerning how wealth should be created and distributed. From still another set of perspectives, romanticism and the other counter material culture movements have bemoaned the rise of the material culture happening at all. In their eyes, excessive desiring of wealth is the root of greed and trampling on the dignity of the poor.

By 2050, the wealth-making and distributing playing field will have been upended yet again. The combination of manufacturing and service automation (as well as actually achieving stable TES) is going to change all the wealth rules once again.

Rights and Rewards as Wealth Categories

One of the upheavals that comes with establishing a TES is that wealth is divided into two broad categories: rights wealth and rewards wealth. (Elsewhere, I call these necessity wealth and luxury wealth.) In the 2010s context, think of the differences between buying food with food stamps, buying a nice car with money made from a well-paying job, and getting perks such as flying around in Air Force One.

This difference between rights wealth and rewards wealth will be even sharper in 2050. And the dividing line will remain an important issue to community members. In those communities that enthusiastically endorse TES, the percentage of wealth distributed in the rights category will grow. In 2050, thanks to constantly growing productivity, communities will be able to support more wealth being put in the rights category.

But people will still aspire to own luxury goods, so rewards wealth will grow as well, and the desire to acquire it will remain strong. One of the big challenges of 2050 is deciding which activities humans should get paid reward wealth for.

What Should Become Rights Wealth?

What things should people of the community be entitled to? What is sufficient food, lodging, and health care? What else should become an entitlement?

2010s and earlier versions of TES have failed. When much of the wealth moved into the rights category, the productivity of the community suffered. The community grew poorer, not wealthier. Notable examples include Greece, Soviet Union, Cuba, North Korea, and pre-Thatcher England. This impoverishment happened because the productivity of the community declined. As the TES became more widespread and comprehensive, fewer people in the community were busy at jobs that created lots of wealth (the difference between working in a factory and regulating one). (Some places, such as Denmark and Sweden, have had better success with handling rights wealth, but there the rights styles and levels of spending have been carefully limited. Community members have cooperated well on these issues.)

The difference that 2050 brings to this mix is that productivity is no longer in the hands of humans. Rather, it is in the hands of cyber. This means that it doesn't matter much how much human activity ends up in nonproductive categories, such as being regulators or doing make work, instead of in productive work categories. The community's wealth no longer depends on human productivity. This means that TES variants with wildly unproductive humans can be sustained as long as those humans don't somehow muck up the cyber side of manufacturing, servicing, and distributing of rights goods and services. (This can happen. This is a real-world threat of the 2050s.)

The spooky side of this is that harsh reality will no longer rein in enthusiastic emotion-based interest groups. TES can become playgrounds for charismatic demagogues who prey on human heart-thinking. ("There are evil secret agents amongst us who will hurt us badly if they get their way. We must be vigilant and root them out before that happens," or "We are going to run this community the way God tells us to run it.")

Communities based on these styles of thinking will be stable, even though they are a long, long way from the US's Founding Fathers' thinking, which was solidly based on the harsh reality of diligently researching how to be productive in colonial North America—a place with lots of wilderness, lots of exploitable resources, and lots of deadly surprises.

What Should Become Rewards Wealth?

As TES grows and rights wealth grows, what should rewards wealth consist of? What activities should rewards wealth reward? Rewards wealth is going to buy different kinds of things from rights wealth. It is going to buy luxuries and perks.

Supporting hobbies will be one form of luxury spending. Closely related is supporting dilettante activities such as preparing superior meals, becoming highly proficient at sports, and vacationing to exotic places.

Exclusivity will be one of the luxuries of 2050. This emotion isn't going to change, but what will be considered exclusive will. What is considered exclusive is partly a fashion call, so there will be surprises as well as constants. Demand for exclusive real estate will be a constant, and wearable accessories will be fashionable.

Here are some thoughts on how rewards wealth will be handed out.

Ways of Rewarding
(Who Gets to be a Kleptocrat?)

In Neolithic village environments, there is little wealth of any kind to hand out because the environment is seminomadic. In Agricultural Age environments, most rewards wealth has been handed out based on social standing in the community—the noble/priest classes getting the most—the few exceptions include occupations such as merchants. An innovation of Industrial Age capitalism was to include productivity as part of the basis for social standing. Capitalism created powerful business classes that at first coexisted alongside the hereditary landed and noble classes and with time displaced them. As the Industrial Revolution evolved even more, those involved in lawmaking and enforcing have been added to the wealthy classes: regulators, bureaucrats, and politicians.

In 2050, humans will still be doing business, but it will mostly be artisanal business, not foundations-of-productivity business (big business). Cyber will be handling that. This means that improving productivity will lose standing as a criterion for handing out reward wealth, and the criteria move back to centering on social standing as it did in the Agricultural Age.

Ways of rewarding humans will still be a chronic issue in 2050 communities, and it will be something humans have a lot of say in, no matter how divorced they are from the harsh realities of productivity. The following are some examples of how instinctive thinking can shape how rewards wealth is handed out.

Leadership

As has been true throughout the ages, people will reward other people for being leaders. A person who commands respect and directs many other people in an activity will be rewarded with luxury money.

There will be many kinds of leaders and many kinds of leadership styles. The most desirable leaders are those who gather the community to accomplish big goals. JFK's "Ask not what your country can do for you—ask what you can do for your country," personifies this big vision leadership style.

Another common style of leader is one who stokes community fears. Given how divorced the TES-acclimated person will be from dealing with material harsh realities, emotions can run strongly and be the source for much decision making. In such an atmosphere, stoking fears will work quite well to produce leaders, and they will be well-rewarded. An example of ruthless leadership in the 2010s is Russia's Vladimir Putin.

Entertainment

Entertainment is something that humans will spend even more time on than they do in the 2010s, and success will be rewarded much the same way. There will be a whole lot of nearly anonymous aspirants and a few high-profile rock stars. The rock stars will be highly rewarded.

Entertainment has part of its roots in Neolithic village environment begging. It split from begging much as religion and science split from fortune telling. It has grown in social importance as prosperity has grown and will continue to do so. It will keep changing form to suit existing local tastes and conditions.

As productivity loses importance, entertainment will become even more important. In 2050, it will become what humans spend most of their time, attention, and effort on. Community members will spend more and more time on both sides of the entertainment equation—producers and audience members.

Good Deeds

People like seeing good deeds happen. Given that, one style of activity that can be rewarded with luxury money is for Good Samaritan and pay-it-forward type activities. This could include things that are familiar today such as running charities, and it could include things such as winning prizes for other good works the media and public choose to recognize (Nobel Prizes). These contests and prizes could become substantial parts of the reward system.

What are considered good deeds will be fashionable: The choices will change quickly with time, location, and the decider. As a result, there will be arguments about how good a deed really is.

41

Gaming the System

People love to be rewarded for gaming the system. If someone is playing a loophole or sneaking stuff out behind someone else's back, succeeding at this feels good. As a result, letting people game the system buys a lot of social peace. While the TES community members may say they don't want this, and the media may amplify this sound and fury, the social peace it brings can be substantial.

System gaming can be supported by cyber-lying. When cyber can support hypocrisy, system gaming can be structured to keep the expense low and the social reward high. As a side benefit, discovery of system gaming can also be a supported activity. This also brings a lot of good feelings; people enjoy cheaters getting their just desserts. The chattering classes will love this in 2050 as much as they do in the 2010s.

How this discovery and punishment process will be handled will be highly variable among communities. A possible surprise use of the technology can be that most of these discoveries are in reality fabrications whipped up by cyber simply for human consumption. The perpetrators being shown in news videos are not real people; this is Big Brother newspeak stuff.

Personal Industry

Some people will still want to work for a living. This, too, is a deep human instinct. The reward for good work attitude thrived in the Industrial Age as indicated by the rise of Protestantism. What kind of work is good work and how to reward it will be one of the most variable parts of the 2050 environment. One style will be doing artisanal work that creates high demand.

One ongoing emotional conflict will be the emotional gap between the system gamers and the hard workers. Being rewarded for hard work conflicts with the wish to be rewarded for doing no work; this emotional difference will still be a hot point in 2050.

Related to the above, how will politicians and bureaucrats fit into this system? In part, much of the bureaucracy side can be taken over by cyber. Once the rules are established, cyber can help people hoop-jump as well as humans can—if not better. However, how much of this cyber handling hoop-jumping is actually implemented will be highly variable.

What will politicians still control? What laws can they make? The craziness of the Congressmen as depicted by their opponents in 2015 can be a model for 2050. These legislators will talk, but it is just cheap talk as far as cyber-controlled activities are concerned. What will legislators and governors still control? Will they do more than just think of new hoops for humans to jump through? How damaging to the cyber-managed systems can they be? Will they send cops after totally automated factories and tell them to shut down? (For more on this, see Saving and Investing)

Gate Keeping

Related to hard work is gate keeping. These are the regulators who want to see that things such as fairness and safety are just as high priority as getting stuff made and getting innovative ideas brought to reality. This feeling is powered by the prescriptionist instinct. There will be lots of people who want to be regulators, and they will have specific right ways in mind as they do their inspecting of other people's activities. The list of what people want to prescribe is nearly endless. The question each community will face is how to balance the prescription instinct against the creating of opportunities to experiment with many different ways of doing things that prosperity offers.

Mixed with this will be the disconnect from the harsh reality that cyber is providing. A surprise use of cyber human manipulating abilities is that they may be able to find inventive ways to lie that allow both prescriptionists and experimenters to have their way. What the regulators really regulate may have little to do with what the experimenters experiment with.

Plight

Begging—rewarding for plight—goes way, way back. Because of its deep emotional roots, it is not going to stop in the TES environments, which is ironic. As it always has, it will change form to suit existing conditions because it is very much an advertising activity. And, as it always has, it will annoy many people as well as salve many people.

Saving and Investing

Saving and investing are icons of human activities successfully adapting to the Industrial Age environment. These become big and important activities only when human activities get complex and there is lots of wealth to be made and passed around. These are comparatively new human activities, and they are going to be changed a lot by the time 2050 rolls around.

As cyber takes over big business, it will also take over making the choices of what to invest in. It will decide what new machines to get, what the machines should produce, and what new services to provide. This cyber takeover of big business investing has to happen because so few humans will understand what is happening in the world of cyber manufacturing and services.

This means that capitalism—the act of humans investing in companies—will once again change dramatically. What a human board of directors will do for a company where cyber is making all the investing decisions is not clear.

Because the role of human investors will change so dramatically, human investing will change. There will still be financial instruments for humans to invest in, but their forms will be quite different from the investing instruments of the 2010s. They will be even more abstract and disconnected from the day-to-day activities of wealth generation than those of the 2010s.

One of the instincts that powers investing is the gambling instinct. This element is likely to endure, and it will be something that disconnects human investing from cyber investing. The human side will continue to support the thrill of gambling even when the investing process becomes a lot more certain. For this reason, human investing in smaller-scale, artisanal activities is likely to thrive.

Saving and investing are tightly tied to luxury wealth. It may be so tightly tied that rights wealth can't be used for these activities.

Unanswered questions: How will humans be innovators in this cyber-dominated big business world? Where will they hold out and still be useful? In these hold-out areas, how can they invest in their dreams? How can they be educated to do these kinds of innovating and investing activities?

Being Fair

The question of how to distribute luxury wealth fairly will be a hot issue. It will rage—and rage strongly. So will the related question of what are acceptable working conditions. Both are deeply emotional issues with lots of instinctive thinking, and both suffer from the curse of being important. Lots of people have strong opinions on what the right way to handle these issues is and feel those opinions should be respected.

This will be one of the areas in which the different TES communities vary widely in what they choose to be the right formulas.

How cyber (in particular cyber lying) will affect these issues is unclear. There will be surprises.

Further Reading

This February 23, 2013, *New York Times* op-ed, "A World Without Work," by Ross Douthat talks about how unemployment is starting at the bottom and working its way up. That is the opposite of what was expected of a prosperous utopia by most visionaries.

I Am A Rock

"Okay. You set a record. Now let's go out and celebrate."

This is Nancy-672 talking, Simon's muse who is in cyberspace. She is talking to him through his communicator. Simon had just completed Level 1001 on EPICGTA v34. He had been at it for nine straight hours. Well, that included the walk-around-the-room breaks that Nancy prodded him to take every thirty minutes as well as lunch and snack breaks. But any way you counted it, the sun had moved from sunrise to sunset (not that Simon saw it), and now Simon's hands are aching.

"Go out? To where?" he says as he stands up and stretches a deep, satisfying stretch.

"Well, you can meet Jamie and Arthur. They're at Starscents."

"They'll be jealous, not happy." He waves off that idea. He pounds his chest and says, "Man! I feel like I should be able to add to my tattoo! When does the next installment on that come up?"

"In two weeks."

"Can I pay to get it now? I'm feeling so hot!"

"What will you pay with?"

"Fuck! Don't bring harsh reality into this conversation. You bring her in, and you're asking for a bitch-slap."

"Sorry."

"Okay … no tattoo for now. What can I do?"

"How about something to eat? You feel hungry."

"I don't want any more Nutribuzz."

"You don't want to save the world?"

"I've saved the world enough today. I made it to level 1001. Time for celebrating with some real food, something that was growing in the ground."

"Good! That will get you outside and walking around. That's good for you. But keep in mind it's snowing now."

"Shit! Really?"

"It happens."

"Okay. I guess the good news is the street won't be crowded."

Simon is twenty years old. Given how much time he spends in the basement, he looks pretty decent, with a slim body and dark hair. His tattoos haven't gotten too outlandish yet. He puts on some street-suitable clothes, adds a jacket and hat for the cold weather, and heads out. The street is far from completely empty. The trucks and taxis are buzzing up and down, and there are plenty of gentrifiers walking around showing off their luxury fashions and arm-candy muses.

"What wasters," mutters Simon, but he says it because he is wishing he could do some more serious wasting himself. He'd particularly like to be wasting time talking with a real interesting girl, a real one, but those are even fewer and farther between than interesting guys. So instead he has Nancy, his necessity muse, and he talks with her a lot over his communicator. One day, she may turn into a real, stand-beside-him muse, but like fancier tattooing, that takes luxury money.

As he walks, she says, "There's a poetry reading in an hour at Gafongle's."

"That's cheap. Does that count toward saving the world?"

"If it grows your soul, it sure does."

"Can I afford it?"

"Silly boy!"

Simon heads for Gafongle's. Gafongle's has a necessity food menu as well as a luxury food menu, so Simon can get a meal there using his necessity money and then stay for the entertainment. Poetry reading is not quite as get-down-home-style as pop-and-country karaoke, but it is sure no symphony concert, and he hasn't won a free ticket to any hot-item rock or rave concerts for a month.

As he walks to Gafongle's, he mutters, "Man! I should be doing something to make some more luxury money. I've got the time these days."

"Why don't you do some more tour guiding at the art museum?"

"Man! That's volunteer work. That's resume building, not getting me luxury money!"

There is a pause while Nancy scans online information. She says, "There's a salesperson opening at the Newton Gallery."

"That place? Geez, I thought you wanted me growing my soul, not selling it."

"Isn't that a common tradeoff for some luxury?"

Simon thinks about it as he walks and says, "What can I do that I have passion about … and get some luxury money?"

Nancy scans some more. Nothing comes up before Simon gets to Gafongle's.

<<<*>>>

In Gafongle's Simon loads up his plate at the necessity buffet and sits down at an empty table at the back of the room. Eating is going to keep him happy, so being far from the stage isn't bothering him.

Simon likes poetry. He writes some himself, and he's been on stage here twice. It is a quiet Tuesday night, so the luxury tables are nearly as empty as the necessity tables.

Al Gafongle, the owner, is cruising the place. He is networking with customers. He comes by Simon's table and joins him for a bit.

"Stormy tonight. Inspiring, perhaps?" he says to Simon.

"Well, I did just reach level 1001 on EPICGTA."

Al thinks and then says, "Yeah, I guess that counts. Congrats. Any verse coming out of that?"

"Not so far." Simon keeps munching away.

"Well, keep at it … one, or the other, or both." He moves on.

<<<*>>>

As he is walking back, the snow is falling much harder and the wind is blowing in his face.

"Fuck," he mutters. He can't zombie out. He has to pay attention to where he is walking.

"You could raise a child," Nancy says to him.

"What! Get married … to a girl?"

"No. Raise a necessity child. There's lots of luxury money in that these days."

"Whew … I can think about that," says Simon and then he is back to concentrating on his walking.

<<<*>>>

Safe within his room, he does some more research on this baby-raising idea.

"It sure does pay well. Why is that?" he asks Nancy.

"Supply and demand. There aren't many humans willing to get into this, but there are a lot of necessity children being created these days."

"Why is that?"

Nancy pauses before she answers. "A couple reasons: The first is that many necessity kids are raised to live in non-Earth-surface environments— places like high-pressure undersea areas and low pressure/low gravity space stations. They won't ever live on Earth's surface. Thanks to new technology, we are figuring out a lot more ways to exploit these hostile-to-human environments. But a lot of regular humans find these necessity humans too strange to deal with. They stay away, far away.

"The second is that too many humans are like you: They aren't willing to commit to a family raising project. There aren't enough regular humans being raised to sustain the population. Some places see this as a crisis needing fixing, and necessity children are being raised as regular humans to fill these places up."

Simon snorts, "That second one is not surprising, given how risky a commitment is in the legal sense. You can really get hung out to dry if things go sour … especially men these days. I've researched that! This 'We worry that you may become a deadbeat dad' baloney can screw up your social and financial life for decades. And who wants to spend years changing diapers, anyway?"

"You won't be changing diapers. The cyber do that just fine. And these are elementary school-aged kids."

He considers this some more and says, "Am I going to get into that 'You're a Terrible Dad Now' kind of issue if this necessity child raising goes sour?"

"Not if you work with the off-surface ones ... the really strange ones. The rules are quite different. Necessity kids who don't live on the surface are so new and strange that the old instincts haven't taken hold—not much, anyway. That's the kind I recommend for you."

Simon does some more research. "Neat! I can do this from my room." After some more research, he says, "There's a lot I have to learn."

Nancy giggles a little at that.

<<<*>>>

Jamie and Arthur stare in amazement and say, "You're going to do what?"

Simon giggles a little at that. "I'm going to be grandma to some green-skinned, four-eyed baby. Well ... I don't know what it's going to look like, and I won't be changing its diapers. But I will be raising it. And I'll get some good luxury money for doing it."

Jamie takes another swig of organic, vat-grown beer, "It sounds crazy. Why did they choose you?"

"I did some training, and they say I have good patience and perseverance. My level 1001 helped. The examiner said that getting that showed those virtues."

Arthur says, "Oh yeah. You did get that. Are your hands still aching?" He says this last part in an almost hopeful way.

Simon grins back at that. "I'm at 1010 now. But I've had to slow down a lot. This baby-tending training takes a lot of time and attention."

"I hear that babies do too," muses Jamie.

<<<*>>>

Six months later, Simon is certified and ready to take on his first bundle of joy. He is on the screen with Elaine Banner, the administrator in charge of this group of children.

She says, "Hi, Simon. So you're ready for your child? Children, actually."

"Yup. I'm certified, and I have my avatar suit here in the room with me. I'm checked out on that too, and it's checked out and ready to roll."

"You are going to be caring for a class of porpoise kids. They are being raised to tend farms in coral reefs; they will be living in the Great Barrier Reef off of Australia."

"I've never heard of this. What do they grow there?"

"Oh, things like Yalmamma, but that's not your primary concern. They will be learning those skills from marine agronomists. You're there to teach them socializing, both with each other and land-based humans."

"Wow! Have they got gills?"

"They are porpoise kids, not shark kids. They breathe air, but they acclimate early on to wearing breathing kits. No, the big difference is their air is at ten atmospheres. You'd get sick quickly if you were in the room with them."

"Okay. I'm ready to suit up."

"Good. I'm linking your suit to the avatar in their classroom. Have fun."

<<<*>>>

At the end of the session, Simon climbs out of his suit and grins. The session went well. He did well, the kids did well, and he enjoyed the experience. This is feeling as good as getting to level 1001. He is now doing something that he can be passionate about, and it will get him some good luxury money as well.

He pounds his chest and says, "Nancy, if this keeps going this well, you may have a body next year … after I get my tattoos finished."

"You don't want a girlfriend first?"

"Why risk mixing it up with some pig-in-a-poke when I know I'll enjoy you! And enjoy you a lot. Mwah!" He blows her a kiss.

Friends and female companionship can come later, much later if necessary.

Today, he is a rock.

Education Systems in 2050

Introduction

Education systems are going to undergo as much change as big business systems. How things are taught will change and what will be taught will change as well. These changes will be adaptations to two big changes in what humans are experiencing in 2050.

- What is important for humans to know when living in the various TES environments.
- The ubiquity of information that is available via cyber systems such as the communications networks and cyber muses.

The overall result is that in 2050 education will be much more dynamic than what we experience in the 2010s. It is going to happen pretty much all the time, it will be mixed with a lot of entertainment, and it will be mixed with a lot of cyber interaction. Attending an education event in a classroom will happen mostly so humans interested in the same topic can come together for some face-to-face interaction and to participate in hands-on projects where many hands help the learning process.

What follows are some details on my speculations of what educating will be like in 2050.

The Ubiquity of Information

Even in the 2010s, people are commenting on how much information is available compared to what previous generations had, how quickly it is available, and how easy it is to sort through. In business circles, this is called

Big Data. This trend is going to continue. By 2050, the communications systems between the human mind and the cyber-communications networks are going to be more sophisticated than they are today, and cyber is going to have a lot more data available.

In science fiction, there are many stories that describe a direct brain-computer network link. I think that by 2050, we will be much closer, but given how hard it is to define consciousness in the brain, there will still be limits. The transition between information contained in the human brain and information in the cyber network will not be as transparent to the human thinker as the way information held in short- and long-term memory in the brain feels to be.

But as this human-cyber link gets faster, easier, and covers more topics, it leads directly to the question: What should our children be learning as they are growing up? What should we be putting in the human side of this system?

Something else to keep in mind is that there will be better and poorer ways to make this brain-network connection. There will be necessity connectivity—which all humans of the community have access to—and many kinds of luxury connectivity—better connections that luxury money can buy. This too will relate strongly to education systems. There will be necessity and luxury education systems as well. This difference will be fueled by instinct as well, the instinct that parents want the best for their children.

What Gets Covered in Necessity Education

The basic goal of necessity education is to put into a child's brain what is needed to function well as an adult in the community. Keep in mind that this community doesn't have humans working in big business occupations as we know them in the 2010s. These big business jobs are being done by cyber. So what will humans be doing? What will they need to learn to do these 2050s-era activities? And most importantly, what do they need to learn that will help them feel deeply enfranchised in their community so they don't cause a lot of waste or damage?

Social skills are going to be a big part of what gets taught. Entertainment skills are likely to be another big part. Because of the disconnect from Industrial Age necessity skills (get-a-job skills of the 2010s), lots of opinion and urban legend are going to be part of the education package as well.

How It Is Taught

With so much communication bandwidth available at such personal levels, having a human teacher stand in front of a whiteboard, talk, and show videos is going to be a real slow and limited way to learn. What will replace this system?

In place of showing videos, the teacher (cyber or human) can suggest communications channels for students to watch. These can be suggested as homework activity. What will take place in class are hands-on projects of various sorts. Examples are as follows.

- STEM: Form a team, research and design a part, have a 3D printer in the classroom make it, attach it, and see how well it works.
- Entertainment: Do stage performances of various sorts. This will be much like the performing arts of the 2010s, and the goals will be much the same. The updates will be the revolutions in recording technologies that happen between now and then.
- Civics/community service: Research how to support a cause. This will be mostly about building enfranchisement. This will be much like what is done in the 2010s.
- In ruthless leader environments propaganda will also be part of the curriculum. This will build enfranchisement.

What Gets Covered in Luxury Education?

Luxury education is for going beyond the basics. Once again, the big issue is what beyond-the-basics activities will consist of for humans. Will they be mostly dilettante activities? If so, then the school courses will be about developing dilettante skills.

What will humans still be doing that is vital to humanity's progress? What will humans be involved in that makes the world a better place on the grand scale (vs a small-scale dilettante)? As these are identified, learning them will be a different form of luxury education from dilettante educating.

Like various kinds of homeschooling does in the 2010s, some kinds of luxury learning in 2050 are likely to run against the grain of some community prescriptionists. This means there will be heated opinions about what is acceptable to teach and what is not. As an example: If someone wants to raise a child to join the 2050 equivalent of military

Special Forces, the raising style engaged in will be quite shocking to Tender Snowflake-style prescriptionists.

How It Is Taught

Luxury teaching will be much more varied than necessity teaching for two reasons. One, the range of topics being taught will be wider. Two, the parents will take pride in how their kids are being taught, which means some fashion instincts will come to play. Some of the teaching styles will be trendy.

Note: The Vol. 92 No. 13 of *The Kiplinger Letter* (March 2015) reports that homeschooling is on the rise and not just for religious reasons. It notes that homeschoolers are testing well, are self-starters, and have a strong work ethic. This indicates that homeschooling-style techniques are likely to be a large part of the luxury education systems of 2050.

Teaching "Necessity Kids"

As a large majority of humans come to live in prosperous urban environments, there won't be enough children born and raised to sustain the human population. Historically, this urban deficit has been made up for by people from poor, rural areas moving to cities. But in 2050, there won't be any poor, rural humans, except for a handful of counterculture humans who set up communes and deliberately live in poverty.

In response, the population can either decline or other sources for creating and raising human children can be created. These other children will be Necessity Children, and they will be raised in environments that are quite different from both TES necessity family environments and luxury family environments. The raising environment will be almost completely cyber-sustained—no human parents involved.

These humans are likely to remain distinct from the other human populations. This will be the 2050s incarnation of separate-but-equal unless some sort of social justice movement kicks in. With this in mind, how they will be taught will also be quite distinctive.

How Teasing, Pranks, and Playgrounds Mix with Ubiquitous Surveillance

Ubiquitous surveillance is coming because it is so helpful in reducing so many kinds of waste. One of the questions that flows from having ubiquitous surveillance is how kids are going to be kids in this environment?

How are they going to play on a playground, how are they going to tease each other, and how are they going to pull pranks? These kinds of activities are part of socializing and growing up. Are they going to be tolerated, or are they going to be replaced with different kinds of activities that are surveillance-friendly?

At this point, I don't know.

What Practices will Instinctive Thinking Continue to Support?

Some things won't change because they are supported by instinctive thinking.

- Some kids will envy other kids for their material ossessions. This means expensive items that can be worn or otherwise displayed will show up in school environments. Unless kids are trained otherwise, there will be excess and abuse surrounding these.
- Cosplay is fun. There will be a lot more of it happening in school environments.
- The struggle between teaching prescription and critical thinking will continue. Related are issues such as what are good science and good history facts (vs pseudoscience and propaganda) will continue to be debated hotly. Instinctive thinking will favor teaching prescription and pseudofacts. Further related is: How valuable will critical thinking be in the TES environment? When people are not affecting harsh reality, critical thinking becomes an unpleasant luxury (unpleasant in the sense that it is hard to learn and causes lots of arguing when it is engaged in).
- In this world of fast and easy cyber linking, what is peer pressure going to be like? Who is the peer a child pays attention to? Is it other children or the child's cyber muse?

Conclusion

Education systems are going to be different in 2050. They will be shaped by changes in what is valuable for people to know and by the fast and friendly linkages people have to the cyber network. What information should be put into a child's brain as well as the social circumstances that the education takes place in will both be very different.

School Daze

Note: This is a luxury education experience being described, not a necessity one.

Troy walks into the classroom in zombie mode. He is not alone; there is a file of boys and girls walking in, and all have their attention on their communicators as their legs take the steps that get them into the classroom and into their seats. He doesn't bother to focus on Eloise Klandis, the teacher, until she flashes a signal to all the communicators in the classroom that she is ready to begin. This is a science and engineering class so the kids are using metric system.

"Class, today we are going to be working on how to make a better seesaw. You did your research yesterday. Now I want each group to design one. We will print out your results on the class 3D printer. That done, we will put them up in the playground and see how they work. Are there any questions? If not, gather in your groups, and let's get started."

Troy is working with Sandy, Jimmy, and Paul. The foursome take about ten minutes to finish their design. Proudly, they wave to Eloise.

Eloise comes over and inspects their work. "Hmm ... a ten-centimeter board. That's going to be big and heavy. It will use a lot of material, and can you carry it out? Check out much it will weigh."

They do, on their communicators. "Fifty-one kilograms," says Troy.

"What weighs fifty kilograms in this room?" Eloise asks.

More research. "Two desks," comes back the answer.

"Go try lifting two desks." The kids do, and she hears them say that it is difficult.

"What can you do about it?"

The kids do more research. Nothing comes to them quickly. After a minute, Eloise says, "I'll give you a hint: foam." This group now begins discovering the virtues of foam cores and composite-fibered exteriors. They get busy redesigning while Eloise moves on to another group.

At recess, the kids carry the printed-out seesaws out to the playground and field test them. One group was inspired by a design they found that was used in Korea. It was low-slung and designed to be jumped on, not sat upon. This was something different and curious to the whole class. Most work out as expected, but there are surprises. For one, most groan when they move. A couple squeak too. Two break, and one proves unusually wobbly and bouncy. The kids are loving that one ... until it breaks, too. Troy and his group ask this other group that made the wobbly one if they can work with their design and see if they can improve it. Eloise hints that is both acceptable and a good idea. The design is shared around, and next week, they will all try again.

After playground time, Troy has a history class. Today, they are studying the Civil War. Kansas Kanab is teaching.

He says, "As you have been told before, the goal of learning history is to learn from the mistakes. To do that, you need to learn what really happened. That is often difficult because to many people, history is a form of editorializing. They have an opinion, and they are going to shape the history story so that it supports their opinion.

"What can you, a learner of history, do to get an accurate story?"

Sandy, sitting next to Troy, raises her hand. "Read and listen to many points of view."

"That's a good start. That will give you many choices. Then what can you do?"

Troy raises his hand. "Read about what else is happening at the time. See if the stories fit together well."

Kansas is impressed. "Very good, Troy! How did you come up with that?"

"My grandfather is a history buff. He loves this stuff."

"That's a blessing, indeed. So … how does that apply to why the Civil War got started?"

"Well, a lot of what I read says the war got started because of slavery. The people in the Northern states all said it was a bad idea and the slaves should be freed, and the people in the Southern states all said it was a good idea and they shouldn't. I've also seen that in a lot movies."

"And …"

"I've read other things, and those other things make a lot more sense. One thing they say is that lots of people in the South didn't like slavery either."

"Good, and what else?"

"There were a lot of economics involved. How to treat cotton was a big one."

"Why was that big?"

"It was big because it was both big business and new. Something called the cotton gin had just been invented by Eli Whitney, and it was spreading around."

Kansas interrupts. "His wasn't the first, but it was the first mechanical one."

"Yeah, and it worked a whole lot better. So much better that growing cotton got real profitable. It was replacing wool as a standard fabric."

"A big change, indeed. So, what does this have to do with the Civil War?"

"Well, according to some of the things I read, the people of the North wanted the South to sell cotton just to them. And the Southern states wanted to sell it to the whole world, France and England in particular."

"And so …"

"And so this was a pocketbook issue as much as a slavery issue—or maybe even more."

"Interesting. Do you have anything else to support this assertion?"

"Yeah. There was a bank panic in 1857. From what I have read, these bank panics always scared a lot of people. Their money was getting messed with in really bad ways. So, add everyone in the country getting scared about money, to the South getting scared the north would cut them out of lucrative cotton markets. People could do some really crazy shi ... er, stuff, and start a war is one of them."

"Nice insight, Troy. This is something you can write up for your paper."

Troy is happy. So is Kansas.

In his geography class, the topic is the Suez Canal.

"Today we are going to do two things: Simulate operating the canal and simulate planning to build an enlargement of the canal so it can carry more ships."

Troy and the other students get deep into zombie mode as they use the online canal simulator to schedule ship traffic and see if they can improve throughput. Once they get the hang of the scheduling, they use that insight to add input to the canal enlargement planner to design ways of increasing the canal capability at low cost.

By the time they are finished, the afternoon is half gone, and it is time for athletics.

Troy is happy to move from brain building to bodybuilding. He is walking down the main school hall to the locker room to get ready for soccer when he sees Kathy Denitch. Nowadays, when he sees her, he gets a new feeling, something he hasn't felt before. He is feeling like she is a pretty girl. Just a year ago, girls were irrelevant to his life, and the romance he saw in movies was icky stuff. But his body is changing, and his thinking is too.

"She's cute," he whispers to himself, and Ace-432, his cyber muse who constantly listens in on his communicator, picks up the whisper.

"Do you want to talk with her?" Ace asks back.

"No!" Troy quickly whispers back, "She's a girl, and I don't have any projects with her."

Ace has nothing to say to that, and Troy goes on to his practice.

At soccer practice, he spends most of the time exercising. He'd love to be playing, but Coach Katzin says building his body is the best way to spend his time on the field. He can play the game at home using his avatar suit. They will play in real life on a real soccer field for the two days before they have a game with another school.

The end of the athletics and the school day comes. He showers and takes a driverless taxi home.

When he gets home, Paris, his older brother by two years, is in his avatar suit in his room. Troy heads for his own room. Once there, he unlimbers his avatar suit and starts playing soccer, mostly to ease his mind. As usual, school has been a busy, hard-thinking day.

After some vigorous soccer playing, it is dinner time. Troy heads down to the living room where he joins Mom, Dad, and Paris at the table. They spend a lot of time eating, a lot of time in zombie mode, and a little time talking with each other.

"What did you study in school today, Troy?" asks Mom.

"We made a seesaw," he answers as he pops out of zombie mode.

"What did you learn?" asks Dad.

"We learned about composite materials. We made our board a fifth of its original weight using a foam core."

"Neat stuff," says Dad, "Knowing that kind of thing will come in handy."

"And we studied the Civil War. Kansas said I was very insightful."

"That's high praise coming from him," says Mom.

"I just saw a good show on the Civil War," says Paris. "It's here," and he sends over to Troy's communicator the link.

"Thanks," says Troy, "I'll catch it later."

After dinner, Troy goes back to his room and works on his secret project: He is building an insect-sized flying drone equipped with a camera. When it is finished, he will send it over to land in a tree outside Kathy Denitch's bedroom window. He may be shy about talking to her face-to-face, but that doesn't mean he doesn't want to see more of her.

He mixes the parts for this in with other school projects he makes on the home 3D printer. His parents and Paris think these parts are just more homework.

The drone is finished and tested. Troy has flown it around the backyard, landed in the tree outside his room, and peered into his window with it.

"Most excellent," he mutters as he looks through his viewer. He checks the time. It is late, but it is perfect for a quick scouting trip to Kathy's house.

He flies it over and gets it landed on the third try. It peers into her window. She is in zombie mode. First, she is talking with friends, but then she starts watching something with a lot of concentration.

"I wonder what she's looking at," mutters Troy.

"I can check," says Ace. A moment later, he says, "She's watching you play soccer at the last school game."

"You lie!" says Troy.

"Well … yes." says Ace. "But she does play with you in avatar games. And she does pay attention when you walk by her in school."

"Really?"

"Really. You two really should say hi."

"Hmm … I'll do that tomorrow."

He flies the drone back home and goes to bed. It has been a really good day, and he has sweet, sweet dreams about tomorrow.

Human Cities and Cyber Cities

Introduction

Humans will continue to live in cities. They will, in fact, migrate to the bigger, more populated cities so there will be fewer mid-sized and small town-sized human clusters.

But midsized and small towns will not disappear. Instead, they will be inhabited by cyber creations who benefit from being close to the factories where manufacturing and service work are done. A town next to an iron mine is an example. This town will have nearly zero human inhabitants, but many creations that support work in the nearby mine.

These two styles of cities are going to function very differently, and that is what this essay is about.

Human-Style Cities

Humans have been urbanizing a lot more since the Industrial Revolution made densely packed humans possible starting in the 1770s. In the last century, the pace has accelerated, and this acceleration will continue. In 2005, the world became fifty percent urban, and by 2050, it will be ninety percent urban.

There is another trend happening. People are clustering into just a few cities, creating very large urban metropolises. (This is similar to the top-forty jobs phenomenon I talk about.) These metropolises will be a mix of gentrified urban living mostly in high-rises in the core, with a mix of different-rise suburban clusters surrounding it. These big metropolises will be supplemented with distant vacation complexes. These will be places like Aspen, Hilton Head, and Disney World are in the 2010s. These will provide vacation locations for the metropolis residents and permanent luxury homes for the more recreation-oriented.

What will not thrive are places that humans have given up on. The most spectacular examples in the 2010s are the inner cities of the Midwest and California, with Detroit and Stockton topping these two lists. These, and many more lower-profile small towns and cities, will lose people. Some will vanish, but many will remain on the map as cyber cities.

The human cities will be mostly about supporting human cultures in various forms. There will be lots of variety as one moves between neighborhoods. An example of this variety is difference in cultures between the various neighborhoods in New York City in the 2010s. This variety means lots of us-vs-them thinking will be supported, and a Hobbiton-style building of road networks will be aspired to in the construction of surrounding suburbs. All-in-all, these neighborhoods will be fun to walk and bike around, but will be killer on getting things done quickly and efficiently when cars and trucks are involved. Delivery drones may make up for some of the inefficiency.

Cyber-Style Cities

In contrast to human cities, cyber-dominated cities are going to be no-nonsense. They will be constantly organized and reorganized to bring maximum productivity to the cyber endeavors conducted in them. This means street layouts will be logical when viewed from the cost-benefit perspective and can be dramatically changed as needed. The likely result is a basic Euclidean grid system with a few tweaks where important structures such as ports and airports necessitate them. The traffic on the streets will move swiftly and efficiently and can reroute swiftly and efficiently when surprises happen.

Likewise, the buildings will be no-nonsense, and they can be built, modified, and torn down quickly as demand changes. Compared to human cities, the cyber cities are going to be kaleidoscopes of change, but they will have very little ornamentation. Even with all this change going on, they will look deadly monotonous to human eyes.

Cyber will dominate the farmlands as well. The big fields and orchards will be farmed by cyber-controlled equipment. The human farmers will be dilettantes who are farming to create luxury foods and fiber. Their farms will be small, idyllic, and few in number and productivity compared to cyber establishments. Most will be urban or suburban; some will be in places with mystic powers such as Mt. Shasta.

How Much Globalization?

One thing that isn't certain when cyber rules big business is how important intricate globe-spanning supply chains will be. These exist today because there are dramatic differences in the cost of production across the regions of the world. These differences exist because of differences in raw material access, infrastructures, cultures, and political boundaries.

In 2050, the cultures will make little difference because cyber will be more homogeneous and humans are no longer much involved. Cyber cities can quickly reorganize so infrastructure differences should not be important. Political boundaries will remain, but what they affect will change. Political boundaries are a human artifice. Cyber won't see any use in them. If people are less interested in and less aware of the nitty-gritty of making and servicing things, political boundaries will affect these processes less and less. This leaves resource-availability differences as the one remaining dramatic difference.

So, in the cyber-dominated world, can much more be crafted and serviced near the end-use point or the critical resource supply point? Will supply chains get much simpler?

Further Reading

This Dec 13, 2014, *The Economist* article, "A troubling trajectory" discusses whether we have hit Peak Trade.

Man and Cyber

This Vision of 2050 is inspired in part by reading Ray Kurzweil's *The Singularity is Near*. In this book, he describes that the growth in computer processing power is proceeding exponentially and will continue to do so for the foreseeable future. The result is that in roughly 2050, it will grow so large that we can't envision what it means. This is the singularity that Kurzweil's book is about.

I agree with Kurzweil that cyber self-awareness will become real, and I agree that the advanced parts will become unintelligible to humans. But I'm an optimist. I believe that there will still be humans, and there will be parts of cyber intelligence/cyber space that humans can meaningfully interact with. My analogy is that cyber and humans will develop a relationship similar to that of humans and cows. (see the Prelude) My vision is that humans will continue to exist, and the cyber will not appear to become oppressive overlords in human eyes any more than humans appear to be oppressive overlords in cows' eyes. Instead, the relationship will be much more subtle and much more comfortable to humans. In humans' eyes, the cyber will be capricious rather than appearing to be dedicated to oppressing humans. They will be much more like ancient Greek gods.

Because of this appearance, human emotion will be to treat them this way. This new trend means that some people will worship them, offer sacrifices to them, and blame natural and social disasters on the cyber gods being angry. This new form of religion will offer a lot of competition to existing religious formats where God provides answers to both personal questions and why various current events are happening.

Cyber and Automation

The worlds of manufacturing, service, and transportation are going to get more and more automated. This means that cyber will be handling these activities. Few humans will be necessary.

An example of the change that this will bring is driverless cars. I envision that as driverless cars become ubiquitous, the role of cars in our society will change dramatically. These days, car ownership and the ability to drive a car are sources of pride. When driverless cars become the norm, who will bother to learn to drive? And if you aren't going to learn to drive, why bother with owning a car? In sum, cars will become taxis, and getting in one to go from place to place will be taken for granted. Call one up, ride in it, and hop out. You go your way, and it goes its way; there is no more thought given to it. Just like a few people continue to own and ride horses today, a few people will continue to own and drive cars, but they will be hobbyists, not mainstream people or activities.

One surprise outcome of this change in man-car relation is the *Fast and Furious* movies. These 2010s movies have the same relationship between man and car that the 1950s Westerns had between man and horse.

One of the big challenges of foreseeing the 2050s is what people will be doing in place of jobs including running transportation networks.

Cyber and Innovation

The areas where humans and cyber will interact creatively the longest are those where innovation and reacting to surprises remain important. Cyber handles routine well. As more and more human activities become routines that cyber can handle, these will get taken over by cyber. Chess playing and then playing Jeopardy! is an example of this kind of cyber advancing.

Humans who get good analytically based educations will help innovate new ways of making and servicing things and help invent completely new kinds of products to be produced. They will also lead research in science frontiers.

Another area where humans will dominate is disaster recovery. When the things cyber normally handle get all messed up and a lot of playing things by ear is called for, humans will come up with innovative answers more quickly than cyber can.

Super Spy Weekend

The Adventure of Tom Sottm and Mary "Mayking" Cullihain

This is a light-hearted tale of a surprise use of cyber muses—in this case, as androids designed to be super spies. Join Mary "Mayking" Cullihain as she has a wild weekend with Tom Sottm, her personal James Bond.

Chapter 1

Tom Sottm opened his eyes for the first time.

Smoothly and quietly, the storage chamber surrounding him unfolded and transformed into something more like a hospital bed. As it did, Mary Mayking came over to carefully examine him. As she did so, her lovely red hair cascaded over her shoulders, and the lightweight nurse gown she wore hissed and crinkled lightly as it moved over her body. As she looked him up and down, Tom could see that her eyes were a wonder. Her makeup enhanced her deep-set eyes and their sapphire blue. Her skin was white and clear like finest china. She looked and moved as though she were a young ballerina. He saw this and was delighted but not at all surprised. Mary was someone he knew well and was very happy to know, even though he had never met her before. After all, he had opened eyes for the first time just moments ago.

Yes, Tom was hot for nurse, plus the passion burned in him to complete his mission successfully. He was going to be alive for seven days at most, and he knew it.

"Are you ready, Tom?" Mary asked him as she finished her examination. The voice was as dainty as her ballerina figure. He nodded.

"Get up, and we both need to get dressed in street clothes," she said and headed for the door leading to corridor. He got up quickly and agilely. The android engineers who designed him knew that he would be wakened in an emergency and that time would be of the essence. He needed only a few minutes to adjust to being animated and that had happened while his storage tank was reviving him.

He didn't follow Mary. Instead, he grabbed her, spun her around, and held her close to him as he felt her wondrously soft and compliant female body. For many seconds, she complied and melted into his arms; her breathing quickened, and she kissed back when he put his lips to her lips. But then he stopped. His sense of duty called loud enough to cool the passion. He released her, except for her hand which he continued to hold gently.

When she regained her own presence, she led them out into the corridor, pointed out a room for him, and said, "Your clothes are in there. Get dressed. I'll be doing the same in the next room down. Meet me there."

Tom did as Mary said. He knew from his programming that she was mission control for what was coming up.

He was not surprised to see that she was still dressing when he walked in. She was human. He was not surprised that she didn't startle when he did so. They both knew that while this crisis was brewing, speed was more important than propriety. He was dressed in a moderately nice business suit and she in a quick-to-slip-on summer dress.

She checked the time and said, "It's four Friday afternoon with the Fourth of July coming Monday. First off, you need to collect some mission essentials. Then we both need to get outfitted for the occasion."

"And what is the occasion?" Tom asked.

"Making a splash at the Big Tex-Mex casino in Macau. We need to attract the attention of a certain Hu Lu Chou, princeling billionaire gambler type, and become his guests as he takes his yacht to Hong Kong. While the cruise is in progress we … you … need to inspect Deck Two and give a general report on the contents."

"Hmm … fancy clothes, flash cash, and a suborbital flight to Macau. Sounds like it's going to be a wild weekend!" He pulled her close to him again. "Have we got some time for a little afternoon delight before we run errands?" He kissed her again.

She kissed back, "The suborbital leaves at eleven tonight. I've already made reservations. Can we catch that?"

"Plenty of time," he said as he slipped her dress off.

<<<*>>>

Meanwhile, in an office cubicle not far away, a young, enthusiastic delivery boy walks in on Ellen De Jeans, a coworker of Mary.

"Hi, there. I'm here about those stale super androids that are getting recycled."

Ellen says in a bored, almost-five-on-Friday way, "Not my issue. You need to take that up with Mary Cullihain. She's in the next cubicle."

"Thanks."

He heads to the next cubicle and announces over the wall. "She's not here."

"Umm ... right. She's taking a long weekend vacation. You can see her next Thursday."

"Fine ... except I've got a problem today."

"What is it?"

"I'm here to pick up ten stiffs, but there are only nine on the shipping dock."

"RTA android, if you please, RTA. Ready to animate." Ellen facepalmed and then remembered something. "You were scheduled to pick those up Tuesday, weren't you?"

The delivery boy grins. "They ain't moving; they're stiffs to me." He winked and said, "And if they start moving, they're zombies, and I start running." He thought it was funny, but Ellen didn't move, and he sobered. "Yeah, I'm running ahead of schedule. If I get this done today, then I don't have to show up at my work until noon on Wednesday."

"Still not my issue. Come back Tuesday."

"You said this Mary isn't going to be in until Thursday."

"Grrrr!" Ellen says to herself, and she hopes this kid suffers through a software upgrade that crashes his favorite video game for the whole weekend.

"Okay, okay. Let me look into this ..." She starts pecking at the keyboard and comes up with an answer.

"That last RTA android is up in an animation room and has been animated for a dry run." She looked up at the delivery boy. "You're not going to see it. You have your full load. We'll get the paperwork straightened out next week."

The delivery boy tipped his hat and headed out. Ellen got ready to head out for a long holiday weekend.

<<<*>>>

Meanwhile, in a plush briefing room one floor up, the following conversation was occurring.

"Thank you for taking time ahead of your busy weekend, Senator Mandel."

"My pleasure, Mr. Harkins. The government is spending billions on this program. I figure if I'm going to serve my constituency well, I'd better know more about where these billions are going. What is your position here?"

"I'm assistant director of the super android section of the CERU unit— Covert Emergency Response Unit. When a crisis calls for a particular kind of delicate, low-profile response, we wake up one of our super androids and send him in to kick ass and take names … but very quietly. If the looming problem looks like it solved itself without outside interference, we've done our job well."

"Can you give me an example?"

"Well, you wouldn't recognize our best work." Harkins grins. "But one of the cases that didn't go so well was that of Abdul Abulbul Amir."

"You mean that terrorist chief no one could touch for ten years who was beheaded by his jealous concubine? Served him right, and it couldn't have happened to a more deserving soul. You really were involved in that?"

"We call that one a five-slash-zero. The five is for mission accomplished with no innocent lives lost and little damage done. The zero is for the thick rumors of our involvement. Much as the public would like to believe otherwise, we aren't into mythmaking." He was still grinning.

Mandel looked impressed. "Show me more, Harkins. Show me more!"

Harkins brought up the big display on the wall in the briefing room. It consisted of many little screens and graphs. He then explained, "These days, the various ERU, Emergency Response Unit, departments handle ninety-nine percent of the nation's emergencies. The old FEMA, Homeland Security, various state organizations, and even the Red Cross and a few other charities were combined into a centralized organization. Our motto is that when disaster looms, ERU is there to keep the loom from becoming gloom.

"ERU is the big group. We are CERU, and we're outside this main organization structure and very low profile. Our parent organization was the CIA. We get called in when there's a disaster in progress that involves bad people and the government wants it handled quietly, not publicly. For example, when terrorists want to advertise their cause with violence, we come in to see that advertising never happens, and they were foiled by what looks like bad luck or internal screw ups."

"I see," said Mandel. "I guess that's why the Amir case gets the zero, eh?"

Harkins nodded. "The billions in expense is because this is an insurance policy. We don't know when our kind of disaster will come, so we have to always be prepared. This means steadily manufacturing a lot of androids. They have limited shelf life, so ninety-five percent of them are simply recycled when that life expires. The remaining are routinely activated for dry runs. This ensures that the infrastructure supporting the androids stays sharp. These missions are as close to real as we can make them, so they can cost tens of millions. A dry run like the Amir case might be to infiltrate a Somali war lord's security team and get one of the chief's bodyguards maneuvered into a compromising situation. We don't reveal we've done so, and we don't take advantage. This is just a dry run, simply to prove that our system is sharp enough to do it when a real crisis arises. About one percent of our activity involves a real crisis."

Harkins then took a moment to look at the big display to digest what it was revealing. The various screens were mostly empty, and the graphs were quiet. He said, "It's just before the holiday weekend so not much is happening, but I do see one dry run being conducted. The super androids have monitors built in so we can watch their actions. Let's see what this one is up to."

Harkins enlarged the screen showing Tom Sottm's activities. From his eye sensors, they saw Tom looking into the face of a ravishing-looking woman Harkins didn't recognize. She was clearly enjoying her position under him. Tom's physical readouts confirmed he was near the peak of having a real good time as well.

Harkins blushed and then reduced the screen again after just a few seconds. Mandel laughed and said, "Dry run, you say. I don't think that run is going to stay dry much longer!"

Harkins finally found some words. "The super androids have to be prepared for all kinds of investigating."

The briefing went on a while longer, and Harkins vowed to have someone look into creating some kind of pornography alert on this main console display.

Chapter 2

The afternoon and evening had been whirlwind for Mary—shopping for attention-getting clothes with an admiring lover offering heart-felt opinions backed with a fully charged Iridium Card, some serious styling followed by flying over the streets of the city to get to the suborbital airport, launching on that, and soaring through space for a three-hour journey while they briefed each other on their background stories and Mr. Chou and his associates.

"This is a precautionary mission," Mary reminded Tom. "We may find pay dirt at the far end, and we may not."

"If we don't, that will be better," said Tom. "It means the world is a better place than we worry about."

Mary was a little surprised at the attitude, but loved it.

On the flight, Mary took one break from being with Tom. She got in the lavatory and texted Ellen back at the agency office. "Check out what's happening with the Tom Sottm android!"

<<<*>>>

Ellen, now getting ready for bed, thought hard about waiting until morning. But then she remembered the delivery boy business and got real curious about what was up. She recognized that Tom Sottm was the name of the missing stiff.

Ellen's first reaction as she picked up on the feed was confusion. Why should she, or Mary, care about the dry run of an android? Why was

Mary texting her about it? She watched for a few moments and muttered, "Yeah, it is a super setting, but …" Then she was curious about this Mary that Tom was looking at, this Mary Mayking. She was a looker and clearly enjoying the looks Tom was giving her. Was that Mary name a coincidence? This Mary that Tom was seeing looked nothing like the Mary Cullihain who worked in the next cubicle over, her friend. That Mary was dowdy, chunky, barely blonde, and middle-aged. She was a worker-bee, not a glamour girl. … Ah ha!

"Mary programs these androids! She programmed that look!" she said to herself. Now she watched with both drop-jaw admiration and pure, green envy. "Go girl!" she muttered to herself. And settled in for some very interesting weekend entertainment.

<<<*>>>

At the airport, Mary and Tom split up. Mary headed for the Big Tex-Mex and got their room there. Even though it was now Saturday afternoon, China time, she settled into the sleep her very human body was now grumpily asking for. She would rendezvous with Tom when evening came. Tom headed for the polo grounds; he had a big impression to start making.

At the polo club, Tom introduced himself as an accomplished amateur who had learned to play in Paraguay and was ready to sport a bit. Mary knew Chou was into polo, and she had done her prep as well as her homework. Tom was custom programmed with polo skills, just as he'd been programmed with his image of Mary as a steamy beauty. This, of course, was transparent to Tom. If you'd asked him about it, he would have said, "Of course," and wondered why you were asking.

He played credibly in the warm ups while Chou was prepping for his game. While that was going on, two unexpected things happened: One of Chou's teammates called in sick, and Tom spectacularly saved one of Chou's horses from a serious injury. With gratitude first on his mind, Chou invited Tom to play on his team. He helped them win that day.

As the game ended with Chou all in smiles, he said to Tom, "You are a good man, Mr. Sottm … Tom. I'm headed to the Big Tex-Mex tonight. Come on over, and I'll show you a good time."

"I'd be delighted," said Tom, and he made his way there separately to rendezvous with Mary and slip in another round of afternoon delight.

Tom was feeling very satisfied himself when he said to Mary, "Good looking as you are, sweetheart, maybe we should let you do the heavy lifting on impressing Chou? He's a princeling with a roving eye. We know that."

Mary stiffened just a bit and then said, "No, sweetie. I'm mission control. The closest I'll get is being arm candy. You're the super here."

"You're sure? I feel like I've met my match." He grinned. "How about a nice full body massage before we fancy up for the ball downstairs?"

"I can get super into that!"

Ellen watched as Mary's thermostat soared.

Chapter 3

The casino was plush. The Big Tex-Mex was brand new and beyond upscale. It was top scale, with an Old American West theme. Tom and Mary walked in looking right at home.

For a few minutes, they wandered the floor, looking, but not gawking, at the bling the Tex-Mex had. They got their picture walking over gold bars mined from the Comstock Lode, standing in front of genuine da Vinci paintings, and shaking hands with a Chuck Norris sim that was so faithful it was hard to tell from a clone.

They spotted Chou at one of the semiprivate gambling tables in the back. He waved them over.

"Care to play a bit? We're doing Old West here, Texas Hold 'Em."

"This isn't what you usually play?" asked Tom.

"No, we usually play baccarat, but we're playing theme day today."

"If you are, how about something truly authentic?"

"What did you have in mind?"

"Five-card draw. That's what was really played in the Old West. Texas Hold 'Em dates from the 1970s. A guy named Amarillo Slim popularized it. It was better suited for casino play."

Chou thought for just a moment and then said, "All for going authentic?" and there were nods of agreement around the table. Tom explained the rules.

Chou announced, "Since this is not on the authorized game list, we'll move to a private room." They moved to the new room one floor up from the casino floor.

As they adjourned, Chou's right-hand man whispered in his ear in Cantonese. Tom caught what he was saying: "There's a gaming commissioner here now."

"Keep him happy here on the main floor," Chou replied as if this were the obvious answer to this problem.

Mary played arm candy for a while, but as the concentration around the table deepened, she excused herself to wander and fully indulge in the wonders of this exclusive mash-up of ultrarich and ultra-ostentatious humanity.

She wandered back by the Chuck Norris sim … only this time it was the real Chuck Norris. "He covers for me when I take a break," he told her. She spotted a dozen faces she recognized, some from intelligence reports and some from *Fortune* and *Forbes* articles. She struck up conversations with a couple who were lower profile. They didn't seem to mind.

On the stage floor, clones of Roy Rogers and Gene Autry were singing, and they were good. Autry sang mostly westerns as Rogers did, but occasionally, he'd slip in some zingers such as "Rudolph the Red-Nosed Reindeer" he'd written for other genres. Those clones were illegal in the United States and even China because of IP disputes. Since Crocker v Monsanto ten years prior, genes were categorized as part of the IP universe.

But this was Macau, and the saying was, "What gets cloned in Macau, stays in Macau." Curiously, where the cloning actually happened was an unsolved mystery, one of interest to many commercial intelligence groups.

She was making the most of her moment, but she dropped it all when she got a signal from Tom—time to get back on mission. She headed back to the room, navigating a bunch of obvious security people on the way. They were so obvious that it made her feel like she were in a spy movie and her antenna were up.

In the room, she found a new guest playing, and tension was high. It was a late stage in this round of gaming. Most around the table were despairing and ready to give up. Three were still at peak attention: Tom, Chou, and the newcomer. Mary's quiet entry did not break the tension.

The chips in the center were piled high, and next to the table was a woman costumed as slave Leia with her hands tied behind her back and a security type holding her there. It looked as though she were part of the stakes and that someone was about to win himself a slave girl. She was

Swedish-looking with a Leia-style hairdo and delightful pouty lips that said, "I'm eighteen years old ... and one day." When she showed mock dismay at her plight, her body moved like a gymnast's.

"And I thought I was having a wild weekend," thought Mary. She noticed that Tom still had a good share of chips in front of him, and she thought further, "What if Tom wins?" She pondered this for only a moment before going back to thinking about the mission.

After a moment, Mary recognized the newcomer. It was Chou's younger brother, Gang Bang, and from her reports, she knew there was a lot of sibling rivalry.

As the betting finished its first round and draw cards were being handed out, a helicopter landed noisily on the roof. It was distracting, the game paused, and annoyance grew. But the noise was followed by a security man hastily coming in and whispering frantically in the older Chou's ear. Mary had to suppress a giggle. Whatever he got told, his eyes bugged out as though he were a movie extra in a *Godzilla* movie watching Godzilla himself rise from Tokyo Bay.

"Momma!" he said in Cantonese to his brother, who then joined him in the Godzilla-is-coming look. The game ended right there. The Chous motioned for their respective security people to collect chips and said to each other, "One flees up, one flees down!" They went into a quick rock-paper-scissors. Older Chou won, but before he could announce his choice, Tom came up, put arms around both to calm them, and said in a slight Austrian accent, "Gentlemen, if you want your allowances to live, follow my plan." Tom quickly whispered his plan, both looked at him with surprise, and then agreed. Both quickly headed out.

Mary was watching, dazed, and so were the other players. This was so out of the blue! As soon as the Chous' chips were off the table Tom ordered, "Play! Deal another round!" To Mary, quickly and totally composed, he said "Mary, please collect my chips." She began scooping chips into her handbag. The man holding Leia untied her.

Mary was about half through scooping up the chips and the cards were half dealt when a matronly Chinese voice came thundering down the hallway. Both speeded up and finished just as the elder Mrs. Chou reached the door. The untied woman decided against a dash for the door and instead hid in a corner behind the security guy. Mrs. Chou stormed into the room.

"Where are my boys?" she barked in Cantonese. Clearly, they would not be happy if she found them there, and the thunder of her tone rocked the rest of the players and guards. This was not a tone anyone wanted to hear from Mrs. Chou! She, not the boys, actually owned this place.

But Tom weathered the storm and walked up to Mrs. Chou, introducing himself. She switched to English and paid him just a little mind. "Where are my boys? They know they aren't supposed to gamble. I'm going to skin their hides!" she said when she finally found some English words to match her fury.

Tom answered smoothly, "They are in the basement conference room. We are having a meeting down there, and when they heard you were here, they suggested I come to you with some questions they couldn't answer … some business questions."

She looked at him most peculiarly now, "Business questions?"

"Yes, we've been doing some planning downstairs. Would you care to join us?"

Mary looked around and thought, "Thank goodness the people around this table have enough sense not to start giggling at this!"

Momma Chou clearly only half believed this, but that half was so happy at hearing this straw of hope that she conceded. "Let me show you the way back, Mr. Sottm," and she headed out as fast as she came in. Tom quickly motioned for Mary to get beside him, and they headed down together.

The planning room was a plush meeting room that could handle about twenty people. Hu Lu and Gang Bang were sitting beside each other at a long table. They were conferring.

Momma walked in and sat at the head of the table. Tom and Mary followed her and sat opposite Hu Lu and Gang Bang. Momma had a sour look on her face. She didn't believe that there had been a meeting here, but she was being polite and hopeful. The boys were scared to death—and hopeful.

Tom started the conversation. He watched Momma closely as he talked. He was cold reading her, searching for a hot-button topic.

"Hu Lu, Gang Bang and I were discussing setting up a … long-term relationship … involving a business deal."

"Not something involving shipping, I hope! Their father loved that, but these boys haven't even seen the inside of his office since they were five."

The boys cringed a little at this.

"No! Not at all. Something they are more familiar with. They have been involved with Tex-Mex as it has gone up, right?"

Momma Chou nodded, and Tom continued. "Casinos are a booming business. My clients are looking to capitalize on a booming business … and do it with partners who have … expertise."

Momma Chou huffed a bit and said, "Well, my boys have more expertise in this than I'd like!" She then became more forgiving, "But it is paying off. And it's something that holds their interest."

Tom said, "Exactly! My clients want to set up another casino here in Macau. This market has just begun to open up, and they want to team up with local expertise."

Hu Lu was getting the idea. He jumped in, "We were thinking of an extension on Tex-Mex"

Momma frowned at that. Hu Lu stopped. Tom picked up in his place.

"Yes, my clients where thinking of a completely new theme and on a new site."

Momma Chou jumped on that. "Not Chinese, I hope! There so much low-grade Chinese theme here already."

Tom, still watching her carefully, said, "No, we were thinking something more exotic. Perhaps Japanese."

Momma didn't cringe at that.

"Historic Japanese … say a samurai theme. That should appeal to Americans and Europeans as well. They know the samurai legends."

Momma's eyes were lighting up.

"We can call it Big Shogun! It can sit nearby here. I believe there's some property close at hand." He looked at the boys, a cue for them to add to this story. "I believe Gang Bang was talking about that when you arrived."

At this point, Gang Bang took over confidently. He loved real estate.

"There are three properties nearby with suitable access, utilities, and easements. I'm researching what we could accomplish in the way of zoning changes."

Momma's eyes started to glaze over at this. Property was not one of her interests. She interrupted. She did not sound angry or accusing now; instead, she was bored and thinking about other issues.

She said, "You four sound like you have much to discuss. I have some issues that need handling right now.

Hu Lu, Gang Bang, let me know how this turns out."

Momma stood, and Tom followed suit. "Good to meet you, Mr. Sottm. I hope you and my boys come up with something productive."

She ignored Mary, figured she was just arm candy, and walked out.

Mary thought to herself, "My goodness, he's smooth!" With more amazement as she realized the implications of what she had just witnessed, she thought, "Oh my, he's smooth! And good! I'm thinking I really, really respect him."

With Momma out of the room and out of earshot, the Chou boys stood up. They wore ear-to-ear grins and were deeply relieved.

Hu Lu said, "You have our deepest gratitude, Mr. Sottm … Tom. That was a close call, indeed!

"Let me ask, do you really have clients interested in a casino?"

Tom grinned back, "Is Santa Claus real?

"No. I was just doing my part to get you two off the hook. You looked like you could use some help."

The Chou brothers grinned at each other.

Gang Bang said, "Let us give you and your companion here a little reward. Would you like to join us on our yacht tomorrow? We are doing a cruise of Hong Kong harbor. Just an informal gathering … a day trip to see the harbor sights."

Tom looked at Mary and then said, "We'd be delighted."

"Be at the dock at sunup."

There were grins and handshakes all around.

<<<*>>>

Tom and Mary had their invitation! Time for some more celebrating … up in their room.

As they cuddled, Mary explained, "Okay. We're near finished now, Tom. Once onboard, you need to scout out the yacht, top to bottom. This yacht is family-owned and very much a trophy, one of the biggest in the Hong Kong region, which means there's a lot of the right hand not knowing what the left is doing going on. We need to know if any of that not knowing concerns us. Specifically, you need to inspect and report on deck two. Give a report on the general contents there."

Tom said, "A dry run, eh? Okay, let's make it smooth ... and interesting. I suggest you take the top and I'll take the bottom ... of the ship, that is." He grinned. "I'll do some more research while you get some sleep." He kissed her lovingly, silently dressed, and headed out.

Chapter 4

Tom came back and gently woke Mary in the predawn darkness. He had been a busy boy while she slept the warm fuzzy sleep of a satisfied woman.

"There's a twist coming," he told her after they were dressed and ready to sally forth. "The yacht security is state-of-the-art. This means I'm going to have to shut down most of my external access and some of my metabolism so that I profile as human, not android."

"Ouch!" Mary said quietly.

"And that means you may have to take up some of the heavy lifting, dear." He kissed her tenderly.

"Double ouch!" she said. "Just keep in mind that I don't do this frontline stuff for a living. I'm a back-room, do-it-in-a-cubicle girl!"

"I will, I will." He kissed her again to assure her, and they headed for the dock.

They were guests, just ordinary guests. They took a simple launch to the yacht, not a helicopter or a hydrofoil. On board, they headed to the main deck and mingled for breakfast. There were about thirty other guests there. Mary was eating this up with a spoon. She met the president of Google Maps, a governor from interior China, and a vice president of Ferrari. As breakfast ended, she continued her mingling and exploring upper decks while Tom slipped away to see what he could discover about the lower decks.

At about ten that morning, all the guests were on board and the yacht slipped her moorings to head for Hong Kong. They were in no rush so they would sail by various local islands for some sightseeing.

At 10:20, Tom caught up with Mary and invited her into a nearby men's room. She was expecting some ambush sex, but instead, Tom talked seriously. "There is something going on here."

"What?"

"I asked around with the crew and found that the lowest two decks are off-limits except to members of Chou Special Security. I couldn't external research because I'm so shut down, but I worked up an opportunity to look through their surveillance system. They're running a cloning lab down there … looks more like a factory than a lab."

"You think that's where the Macau clones are coming from?"

"Quite likely, and it's quite likely that those are prototypes for something more ambitious, which is not nearly so innocent."

Mary was speechless. This certainly had not been on her radar. And it was a lot more threatening to her than that: If this became more than yet another dry run, she could get pilloried. *Christ! I don't need this!* was her first thought. Her second was panic. She really wanted to say to Tom, "What do we do now?" but she was mission control! "Umm, what's your assessment of mission possibilities?" she finally came out with.

Tom thought a bit. "I suggest we wreck the ship and let conventional authorities investigate and discover the shenanigans."

"As in sink it?"

"No, not so dramatic. We'll have it run aground. There will be some damage, but no casualties are needed. The ship can then be rescued by conventional authorities. They will investigate what happened, and in the process, they will search the ship."

"Can we call in the cavalry? Let some other branch of ERU take over from here?"

"Well … we're in Macau. Do you think US emergency agencies will be welcomed to rescue something here? Especially considering this is not any kind of emergency yet?" Mary cringed. She now knew firsthand what being between the devil and the deep blue sea felt like.

"Give me twenty to think about this," she finally said. "Meet me back at the room."

They parted ways. Mary had a whole lot to think about now.

She remembered when she first got the idea for this wild weekend project. She was sitting over lunch with Ellen.

"You know, it's such a shame, such a waste, the way these androids are handled. We spend tons of money creating them and then most go to waste. They are just recycled."

"Well, you can store them only a year, and they live for at most seven days when they are animated."

"Yeah, but they can do so much. You'd think they could animate the stale ones and send them to do seven days of charity work or something."

"You'd think," sympathized Ellen. "But then again, we are a secret project."

"And those dry runs," Mary facepalms and then growls like some colonel in a cheesy World War II war movie. "Your mission, android, is to bring back a toothpick from the dining table of Xi Jinping. ... Sheesh! Even more waste!"

"It's the life we live," said Ellen, and she changed the subject.

But Mary's brain now had a seed in it. "Let's waste a tiny, tiny bit less," she thought.

Mary was highly adept at visual-image programming. She hacked the games she played so she was in control of what the player saw. It got her around the boredom of a constant stream of scantily clad heroines and hulky heroes. She'd have Mickey Mouse leading samurai charges.

"If I'm going to have a wild weekend, I can look like whomever I want in my hero's eyes, and I will. This is something I'm going to treasure for a long time. I may as well look my best in it." Thus, Mary Mayking was conceived.

But now, things were out of hand. This was supposed to be an under-the-radar dry run. It was a weekend run of a stale android with next-to-no staff resource involved, just something to show that an android could be animated and missioned by skeletal staff. This Mary Mayking, mission controller, was clearly an alias ... pfft! Who would care?

But now this mission was threatening to transition from dry to live. This would change everything! There was a crack under Mary's feet that was about to open into a yawning pit.

Tom came back to the room. "Yes, this damage-the-boat plan should work nicely. It will flood the lower decks, so the factory will be disabled and discovered. It's effective."

"Damn!" thought Mary. "If he'd only found some reason for us to slip out at this point."

She asked out loud, "Is this cloning business really of our concern? Can we ignore it? Are we making waves where we don't need to?"

Tom looked at her just a little strangely. "Good question: Yes, it is of our concern. My further research, watching through the surveillance system, indicated this group is under the supervision of Dr. Albert Angstrom."

"Dr. Albert 'Mad Scientist' Angstrom? Of the Interpol top-ten list?"

"One and the same. He is an object of interest on deck two." Tom said this is a most satisfied way. He had completed his main mission most decisively.

Mary screamed inwardly as she calmly said outwardly, "Sounds like you have an appropriate plan for the situation. How do we do it?"

Tom explained, "The captain is touring islands. We can help him get off course and run aground on some local rocks. The Hong Kong Port Authority rescuers can make the discovery of skullduggery."

"How do we get this competent captain to be so idiotic?"

Tom smiled. "Well, it's the pilot manning the steering wheel. That's where you, a beautiful woman, come in, my dear. What makes men go crazier than beauty? You distract the pilot … seriously distract him … on the bridge. I'll be meddling with stuff below deck to flood some ballast compartments. That will increase the ship's draft, and it will scrape across rocks that would normally be too deep to be a problem. Voila! Ship hits rocks, rips the bottom of several compartments out, starts sinking, and the pilot, or whomever takes command in the bridge, runs it aground on the rocks or a nearby beach to save it. The Chou Special Security people are caught flat-footed, and the port authority people get to be heroes."

"Sounds brilliant, except that I'm no Mata Hari."

"You're part of this business and have been for years. You've got eighty percent of the knowledge part." He admired her and kissed her, "And a hundred percent of the physical attraction part."

Mary almost burst into to tears as she heard this, "Tom … Tom … I'm not who you think I am."

Mary pulled her phone out and takes a picture of herself. She showed it to Tom. "Who's this?"

Mary sighed. "This is what I look like to everyone but you. I programmed you to see me … in a different light."

Tom looked back and forth a couple times, which is a whole lot for a super android. "Hmm … this explains why the dresses you bought were size twelve. I didn't think much of it at the time." He thinks some more, but not long. "I … I … I still say we stick with plan A. Beauty is in the heart as much as the eye. You don't have to convince this guy to marry you. He just needs to pay attention for a few minutes while the ship heads into some rocks."

Mary said, "You really think I can do this?"

Tom smiled and kissed her, kissed her hard and long, until she melts in his arms. Even with what he now knew, he still loves her, and she's loving that. "I think so."

<<<*>>>

When the moment was right, Tom got Mary to the bridge. Only the pilot was there. Mary walked in and moved around like a young, curious miss.

"Sorry ma'am, guests are allowed here only with an escort," said the pilot.

Mary smiled at him and started reading him: Will he be attracted to a straight-forward approach or a coy approach? She went for sincere flattery.

"What a wonderful view you have up here! You really steer this ship from way up here? It's so big!" She came over and brushed his arm with her hand. The pilot did not brush her off. Instead, he stood tall at the wheel.

The pilot asked, "Are you some kind of rich heiress?"

Mary smiled and blinked knowingly in response and then said, "Would I be up here unescorted if I weren't? My departed husband was a man of authority. I like men who are take charge and in control."

She had found his hot button. When Tom saw this he headed below. He had some deeply delicate hacking to accomplish to get the ship to crash and make it look like someone else's accident.

Mary stood happily beside the pilot. She was channeling her inner child, proudly standing next to her poppa. With that warm feeling in mind she asked, "How come the captain isn't doing this? It's his ship, right?"

The pilot looked at her and explained things as if she were a child, and didn't mind doing so. "Oh, the captain *runs* the ship. He doesn't *pilot* the ship. That's what we pilots do. We decide where it goes.

For instance, you see those rocks over there? They're called the 'Twin Angels' in Cantonese."

Mary looked. "But I see only one rock?"

"The 'twin' is just to the right, under water, and it's caught more than its share of ships with unwary ... *captains* ... at the wheel." He grinned at that. "That's why you have a pilot when you sail in these waters."

She took a step closer and then looked around the bridge. "I've seen bridges in movies. Where's the thingy with a big handle on it that you push around to indicate full speed ahead?"

The pilot thought for a moment, confused, and then laughed. He was getting into the mood. "That's called the ship's telegraph. It was a way to signal the engine room from the bridge. You only find it on old steam ships ... in movies. Nowadays, we have ..."

Mary was hitting it off with the pilot and much more successfully and faster than she thought possible. She got him away from the wheel and then not facing the wheel as she convinced him that bringing her lips to his would enliven his afternoon considerably. She noticed, but he didn't, that the main control panel went through some kind of reset. Tom's work, for sure. She melted in the pilot's embrace as he got more and more excited. It wouldn't be long now.

And it wasn't. There wasn't much deceleration, but for many seconds, there was an ominous ripping sound coming from the bowels of the ship. The pilot stiffened and almost threw Mary off of him.

"Oh, shit!" he said in Cantonese. Then in English, he said, "Get out of here, quickly!" He then mumbled more in Cantonese. He paid no more attention to Mary as he started looking over the main console to assess what had happened. Mary didn't need to be told twice.

Chapter 5

The TV news in Hong Kong that day was abuzz about a yacht crash. A live-action view from a helicopter showed the Chou yacht facing a beach on a small island. It had beached there to avoid sinking. A couple of launches were approaching and departing, hauling off passengers. A helicopter was also departing from the yacht helipad.

Behind the helicopter scene was a Hong Kong announcer explaining, "In local news today, the Chou family yacht experienced a near tragedy. It ran over the infamous Twin Angels rocks and suffered severe damage to its hull. Fortunately, the pilot was able to keep control and beach the craft before too much water was taken on board. The ship is safe, and as we watch, the passengers are being evacuated."

The scene then cut to the helipad next to a public dock in Hong Kong. The VIP passengers were getting off the helicopter. Two, Arnold Tramp and his current arm candy, were being interviewed. They look concerned but otherwise unharmed.

The reporter asked, "Mr. Tramp, what did you experience out there?"

Tramp replied, "It was scary. We were outside admiring the view after finishing breakfast when we heard this god-awful ripping and scraping coming from below us.

"We started heading for the lifeboats, but the staff informed us that would be unnecessary. However, we should brace ourselves for the beaching. That was like a car driving up a snow bank.

"The ship was solid after that, although I understand the lower decks are now flooded. Twenty minutes later, we were on the helicopter headed here. I want to commend the staff, they were very professional."

The reporter then turned to the camera and said, "Thank you, Mr. Tramp. "Preliminary reports indicate that this was a one-in-a-million accident. The pilot suffered a mild epileptic seizure just as the ship's navigation system had a glitch and reset itself. It's not anticipated that safety procedures will have to be reviewed."

While that excitement was going on at the helipad, a launch docked at a private dock. Tom and Mary as well as other guests who wanted to maintain a low profile departed from the launch and started walking up the dock toward the city. Tom and Mary had changed from ostentatious clothes to those that would help them blend in with the crowd.

Their mission accomplished, Tom and Mary now walked away into the street crowd. Within a few steps, they were just part of the crowd, blending in.

Nearby, Dr. Angstrom was attempting the same blending, but he was quickly surrounded by a knot of Interpol officers who had some questions for him. That was nice to see, but what mattered now to Tom and Mary was that the mission was accomplished. It was time for them to quietly disappear. They caught a cab and took it to Victoria Peak Garden overlooking Hong Kong's harbor. There they found a secluded spot and held each other.

"You were wonderful, Tom, so much more than I imagined, and I imagined a lot!" There were tears in Mary's eyes.

Sounding a tiny bit like John Wayne, Tom said, "I was doing my job, Mary. I'm so happy it has turned out this well."

Mary leaned into him and sobbed, "And now you get recycled!"

He stroked her hair calmly, "This is what I was created for. I could hope for no more." After a pause, he said, "There's a recycling center here in Hong Kong. We ... I should go there. I don't think you want to."

She couldn't help herself. "It's just not fair," she sobbed. She looked in his eyes again, those wonderful, understanding eyes. She saw a bit of discomfort in them.

"I'm not human, Mary. Please remember that. I was built to live a week, not a century. I don't aspire to living forever. I aspire to doing my duty, and I've done that and done it well.

"My discomfort is … I love you so, Mary!" he blurted. Calm again, he said, "It may have been your programming, but I think there's more to it than that. You are a wonderful woman.

"In another life, in another era, fate would have made you a gentleman's wife, and he would have been a very happy man. I'm unhappy that that love can't go on forever, but I'm done. I'm comfortable with that. Again, I'm not human. I'm super android."

Mary had no answer to that. Finally, taking all her will to say it, she said, "Leave. I will find my way home."

Tom did.

When she could open her eyes again, Mary looked at the view. She watched the other lovers going hand-in-hand through the park, sneaking touches, cuddles and kisses. She stayed an hour. Sometimes, tears flowed; other times, she kept them in check.

There was nothing else she could do about Tom, and her very human body was reminding her that she had other needs to attend to. She got up and found her way home, moving around now as one of the crowd.

Chapter 6

It was a week later when the three-person review panel got to Mary's incident. The reviewers consisted of Mary's boss, Assistant Director Harkins; Agency Director Munro; and Jones, another assistant director.

Director Munro said, "Mary, this is preliminary. We are reviewing what happened in this incident. There are some oddities about this. It's not quite routine, as I'm sure you're aware. Depending on what we conclude, this incident can disappear into archive oblivion, even with its oddities, or it can become the beginning of a more formal review process. Do you understand?"

"Yes," she spoke quietly.

"Mr. Harkins, I believe you wished to open this inquiry."

Harkins huffed and then began, "Ms Cullihain, You initiated this dry run, which is within your authority when higher authorities are not available. But then you escorted the android on the mission. Unheard of! What did you have in mind? Some kind of wild weekend with your own personal James Bond?"

Mary cringed under all this bluster. She had a defense worked out long before she started this escapade, but her heart wasn't into it now. She was still heartbroken.

"You and that android spent roughly a million of agency resource on fancy clothes, fast transportation, and putting on a show at that casino. And, according to the recordings, about fifteen percent of mission time was spent under cover, shall I put it?"

She cringed again.

"Ms Cullihain, you have gamed the system. What happens if even one in ten of the female workers in this agency decide they want their wild weekend? There would be chaos! I see no choice but censure … severe censure." He huffed his finish. Clearly, he felt deeply embarrassed that this had happened on his watch.

As he spoke, Jones looked more and more annoyed. Munro called on him next. "Mr. Jones?"

Why Jones was getting annoyed became clear. "Harkins, covert activity is not always a by-the-recipe activity. Our enemies are tricky folk. They are constantly looking for ways to 'game the system,' as you call it. Our people need to be ready to climb outside the box and kick ass as needed.

"I'm looking at this, and I see a five/five mission! A well-conducted, heads-up mission that came upon the operatives as a complete surprise. What's to complain about?"

Harkins looked bug-eyed at Jones, and Jones looked back defiantly. Mary thought, "I think the Chou brothers rivalry just got surpassed. Oh my! And this happened on my watch!"

After a moment of thinking, Munro concluded the review. "Gentlemen and Ms Cullihain: The way I see it, we can either condemn or commend. And I think for agency morale and standing with ERU in general, a commendation will end this situation much better. I recommend we give Ms. Cullihain a five/five commendation. I also recommend that you, Mr. Harkins, prepare a report on procedures we can implement to keep wild weekends from becoming some kind of agency routine. Meeting adjourned."

<<<*>>>

Ellen met with Mary outside the meeting room. She heard the verdict, and they headed off for lunch. As soon as they were out of earshot, she said, "Yahoo, girly! You showed 'em!"

Mary smiled faintly and nodded. Ellen was expecting a high five.

"What's the matter, sweetie?"

Mary stopped and looked at her. "I loved him, Ellen, I fell deeply in love. I wasn't expecting that." Ellen put her arm around Mary, and they continued their walk.

"Life is full of surprises."

"It was truly a wild, wild weekend. It was everything I planned for. How could my feelings about it go so wrong? I had my wild weekend with James Bond. But where do I go from here? I'm feeling like there's no tomorrow for me. Gag! I sure don't want a sequel!"

Ellen turned Mary to face her. "Why not, sugar? You've had the perfect romance, just not an endless romance." She gave Mary an encouraging hug.

"You're not doing Tom justice with all this moping around. Do you think he'd like to see you this way? He wanted to make you happy, sugar. He wanted to inspire you, not turn you into some kind of mope who can't do a thing without her man by her!"

Mary thought about this, gave Ellen a hug back, and smiled weakly. "You're right. You're being a true friend, Ellen."

"Damn straight! And I'm going to give you one more piece of advice: Capitalize on this! You've proven a super way to use super androids. Here's what I suggest: You get in touch with those nerdy engineer types at Chou Special Security."

Mary looked shocked at this.

"Not the evil leaders! The I'm-doing-my-job types they hired. That ring is broken now. They'll be looking for work. What would be better than developing wild weekend androids? After all, sweetie, the clones made in Macau stay in Macau."

Mary laughed at that, and the girls now headed for lunch with some new big dreams in mind.

Cyber Muses

Cyber muse is an up-and-coming relationship between computers and humans. As cyber becomes more intelligent, one of its uses will be to stroke human emotions. This will be like video gaming, only with more sophisticated goals than simple entertainment. Creating cyber muses will be an evolving process. Cyber muses will first be designed to interact with human emotions because these are relatively simple forms of human thinking, so it will be easy to design cyber that can interact efficiently with them.

As this interacting skill is mastered, the next question becomes what the invention should be used for? This is a general question that applies to all kinds of inventing. Answering this question will typically talk first about a commodity use (as I call it) and then about surprise uses (as I call them). The commodity use for cyber muses is where they will help a human do something the human already does, but faster, better, and cheaper. Using video games as an example, they help humans stay entertained faster, longer, and cheaper than, say, solitaire and crossword puzzles, and more elaborate alternatives such as nightclubbing and concert-going.

Commodity uses are important, but any invention that becomes noteworthy will then develop surprise uses. These second uses can be described as, "Eh? You can do that with it too? Neat!" These surprise uses are much more interesting. These are the game changers. Cyber muses that do more than simply entertain will be a surprise use.

The name "cyber muse" comes from the expression "Behind every great man there's a good woman." The beginning inspiration was knowledge that women can inspire men to do great things. If a cyber could provide the same kind of inspiration faster, better and cheaper, this would be a wonderful benefit to mankind. My definition of a cyber muse covers many more kinds of human thinking stroking, and most are a lot simpler. Inspiration stroking will be one of the highly advanced forms and will come late in the development cycle.

The first cyber muses will stroke simpler and blunter emotions. The very first conventional ones will simply do video game kinds of entertaining more elaborately than video games do. Right along with them will come those that are enhanced porn. These will stroke sexual emotions. For delight, or for outrage, cyber sex toys will be some of the parade leaders. (This is an example of a commodity use—getting sex stimulation and satisfaction faster, better, and cheaper.)

Coming out in parallel with sex toys will be comfort toys—cute-kitten cyber muses. These will comfort those who are sick, those who are lonely, and others who need care (such as the elderly and infirm). These cyber muses will take the place of cats, comfort pets, and therapists. For those who are just lonely, not sick and lonely, these muses will take the place of visiting tea houses and Starbucks just to see other people.

Following shortly behind the comfort muses in development will be arm-candy cyber, those that are designed to comfort their owner by being on display. These muses are stroking the "keeping up with the Joneses" emotion. This is comforting for the aspiring Donald Trumps of the world.

More complex will be muses for those who want to support a cause and rail against The Man. Related will be those that help young adults party hearty—lots of enthusiasm, but no cause needed. These will be hedonistic muses.

And there will be cyber muses for those who want to raise children without the day-to-day tedium involved. This is a more sophisticated version of the cute-kitten muse mentioned above.

Some of the more difficult to design will be those that inspire artistic, science, and entrepreneurial success. The stroking that successfully powers these kinds of thinking are a subtle mix to figure out.

Getting quick, predictable results and avoiding grief will be the two biggest benefits people see for using the cyber muses. The grief can be the monotonous and grinding kind (such as changing diapers) or the surprisingly deep grief and disastrous outcomes that can happen when complex and committed relationships such as marriages go sour. Avoiding the scorned woman pitfalls and child molestation accusations are two places where muses can do a lot to reduce relationship risk, especially for men, while reducing shaming risk will be especially attractive for women.

Building Community Enfranchisement

Cyber muses will start as toys for the wealthy, but as experience is gained in building and programming them, they will become affordable for everyone. This will include all those humans acclimated to the TES environments that will be ubiquitous in 2050. One of the big roles for muses in the TES environment is keeping humans feeling enfranchised, feeling as though they are an important part of their community. Stroking enfranchisement is going to be a moderately tricky muse skill to develop, but it is an important and widely used one once it is developed. This skill will support both the sacred masculine and the sacred feminine (my terms). In so doing, they will keep their humans relatively crime-free and satisfied with their lot in life.

The Origins of Cyber Muses

Computers have had a profound effect on some kinds of human thinking ever since they have become interactive. Reading the results coming off a card punch machine produces limited excitement, but Space War and Doctor/ELIZA, both real-time interactive programs, date to the 1960s, and both could entertain some people for hours on end.

This interest in playing for long time periods indicates they were stroking instinctive elements of human thinking, and this makes them the primitive forbearers of what will be cyber muses in 2050.

In my definition, cyber muses are computers that are designed to stroke human thinking. This is a much broader definition than that of the classic inspirational muse. How the muses stroke thinking will be quite varied. The muses can be purely programs, like today's video games, or they can be inhabiting tangible bodies, which can range from robot-looking (in the contemporary sense) to more and more human looking and feeling (as in, androids) as cyber design skill grows. The classic muse definition form will be a late and sophisticated evolution of the cyber muse concept.

My Muse Experiences

I've had two memorable muse experiences in my life. The first was with my first girlfriend. I met her early in my senior year in high school in 1965. I dated her; we went out for dinners and movies. I took her back to her place, and we'd watch some late-night TV with her sitting in my lap, and then I'd go home. Nothing much sexy happened—she'd slap my hand when I tried to fondle her—but boy, I enjoyed what did happen! She inspired me, and the result was a full letter uptick in my grades for senior year. As a result, I got on the waiting list to get into MIT, but didn't make it … that time.

Then life events intervened. I went off, first to college and then to the Army, and she found another guy her parents liked better. I got a Dear John letter while I was in basic training, and that turned out to be it for that relationship. (She married that other guy a year or two later.) I was without muse, and soon after, I was facing a year in Vietnam. After that, life went on. (And I did get into MIT.)

The second muse came to me in the 2000s while I was teaching in Korea. She was a beautiful, smart woman from Malaysia who wanted to teach English while she was studying at a university in Suwon. She wanted to, I wanted her to, the school owner wanted her to, but the Korean government said no. She wasn't from an approved English-speaking country.

But while that was being decided upon, she and I spent some time together. She was fun to talk to, and she paid attention to what I was saying. My karma soared while I was around her. It said, "If you can hitch up with her, you will have a million babies and they will all be healthy, happy geniuses." I got inspired to do a lot of Baron Rostov writing (see my *Rostov Rising* book). I even wrote a song for one of my stories. Sadly, this relationship did not last more than a few weeks. We talked. I wanted to do more than talk, but she didn't, and I couldn't convince her otherwise. We broke off; it ended quickly and completely; I was again without muse.

These have the following in common.

- In both cases, the muse effect came as a surprise.
- The relationships had exciting potential but did not evolve into something intimate. It was mostly about talking, some dinner and movie-going, and some shared experiences.

- The woman showed a lot of interest in what I had to say.
- One other point of interest: My most memorably romantic relationship did not produce a memorable muse effect. I was sure interested in the woman, and she was sure interested in me, and we did do a lot of cuddling when weekends came. But I don't recall that relationship leading to my producing anything bigger and better in my creative life.

In sum, when the creative muse happens in real life is surprising, and this means that designing a cyber muse that can get consistently impressive results out of "her man" in this way is not going to be easy. On the other hand, designing one that can be part of a heated romance will be a lot easier, so, as pointed out above, this style will likely come first.

Further Reading
- The muses are coming already. This January 20, 2015, *Mail Online* article, "Nearly half of Japanese adults are not having sex: Fatigue and lack of interest blamed as birth rate continues to slide," by Sara Malm, talks about how Japanese men are already doing a lot with virtual girlfriends instead of real ones.
- From YouTube, this is how far we have come already: Japanese Robot Girl Can Do Anything For You!

Looking for Miss Right

Note: In 2050 copyrighting will pick up a new twist. This is explained in the second half of the story.

"Come on, bitch, it's time to get inspired," says Bob Bixby Bucket. He is putting on his best party-hearty clothes. It is Friday night, and he is ready to prance and dance. He will start at Boobies Galore, the hottest, most reasonably priced (in his mind), downtown singles club.

Mary-314, his cyber muse, a luxury model physical one, quietly sighs and silently orders up a driverless taxi. She dresses in a way that will help blend her in nicely with the wallpaper at the club. The taxi comes, and off they go.

In Boobies, Bob and Mary get a seat at a table and quickly order up. Bob starts looking around. He is not discreet; Mary doesn't seem to mind. In only a minute, he spots what he is looking for: the hottest woman in the bar that night for him. She is a blond in a slinky black dress walking from the restroom back to her table. She is also new. She sits with a group of women who are regulars and whom Bob has seen many times before.

"Umm …" says Bob under his breath so only Mary can hear. Mary takes notice of whom he is looking. Bob waves at the ladies. "Hi, ladies. Having fun tonight?" They wave back, but they keep talking with each other. Once in a while, one or another takes a quick look at Bob. Bob waits while Mary now looks hard at the blond.

Mary looks her over carefully and then announces, "She's copyrighted."

Bob sours when he hears that. "Ouch. How come all the good looking ones here are copyrighted?"

"Because that's why they are here—to show off their stuff," responds Mary in a Spockish-logical sort of way.

"Expensive?" asks Bob.

"Yes," replies Mary.

Bob looks around the room some more. He doesn't see anyone else who is both interesting and someone he hasn't looked over before; Mary has investigated them all. This is not what he came to see, and tonight, he is impatient.

"Let's move on," he announces.

He gives a final wave to the ladies, and he and Mary head out to the street.

As they wait for the taxi, Bob says, "Any place you can think of where we can find some cheap ass … er, inspiration?"

She shrugs. She has nothing to offer. "You know what inspires you."

"What's up at the college tonight?"

Mary looks online. "A varsity volleyball game and a musical."

Bob grins at this. "Women's volleyball?"

"Of course."

"A party afterward?"

"Not this time. It's just one game, not part of a tournament."

"A party after the musical?"

"Yes. It's the final performance."

"Okay. Off we go to that."

The taxi comes, and they leave.

<<<*>>>

The musical is a success, and the party afterward is a success for Bob. He spots three hotties.

"There, there, and there," he says quietly to Mary as he points them out. "Any of them copyrighted?"

Mary takes a while to answer. "No." she finally says. Bob brightens at the news. He walks up to the first and starts making conversation with her. Mary is at his side, but she is mousy and unobtrusive. She is, however, listening and watching the girl intently.

After about five minutes of talk, Mary rubs Bob's arm in an unobtrusive way. It is a signal; she has what she needs. They move on to the next lady. He talks with her until Mary signals, then move on to the third. After spending a few minutes with her, Bob and Mary break off and head for a quiet spot.

"Total success?" asks Bob. Mary nods, "I have what I need."

Bob rubs his hands. "Let's blow this place, and you can show me your stuff."

Mary is a little surprised. "You don't want to mix and mingle? This is a nice crowd. I think you would like it a lot better than Boobies."

"Baby, when you strut the right stuff, I like you a whole lot better than Boobies. And I'll like you a whole lot better than here. Spending more time here is risky. You know that. This is college group. Half of these babes are sue-happy feminists."

"Not half, just a couple, and I can help you pick those out. Keep in mind that there are hardly any copyrights here. This group is not the sue-happy type like Boobies is. They have plenty of other things on their minds."

Bob is adamant. "Not interested."

Mary sighs and consents. She calls the taxi, and they head home.

<<<*>>>

As the taxi takes them home, Mary starts changing. She gets cuddly, and the wallpaper look transforms into something a lot more sultry, especially in Bob's eyes. He loves the cuddles and gets real interested in kissing back.

"Which one did you pick?" he asks.

"You can't tell?"

"I don't want to tell. I just want to enjoy, baby, and I am."

Mary cuddles him some more.

Somehow, Bob waits until they are in the apartment before he pins her to the wall and strips off her clothes. She submits, and her breathing and body language says she is loving it. She is like a dancer following his lead, and she responds to every touch.

He grabs a strip of cloth that is hanging by the door and ties her wrists behind her. She waits, face against the wall, arms behind, naked, while he pulls his own clothes off.

"Okay, babe. Let's see who you are. Walk!"

She now walks up the hallway and into the bedroom. Her hips swing, and her hands bounce on her naked behind. She has taken the form of the second college girl, the tall one. Bob's eyes feast, and he follows her into the bedroom.

<<<*>>>

It is now Saturday morning. Bob orders up a simple breakfast and eats it while he catches up on news. Bob left Mary in the bedroom. He expected that she would get up, clean herself, and then sit in a chair and "feature down" because she was off duty.

Instead, she comes in and sits across from him. She is looking plain-Jane Mary. She is not there to seduce him. (Sometimes, she does.) She has something else on her mind.

"You really should get to know those girls from the musical better, Bob," she says in a straightforward way. "I've been researching them, and they are what you're looking for."

"How do you know what I'm looking for?" He is not paying a lot of attention. On Saturday mornings, the news is usually more interesting than Mary, and the rest of the day will be doing something athletic with the guys. Usually, she comes back up on his radar as the sun goes down.

"Because I'm with you looking every weekend, silly. And because I'm your muse, I know you."

Now he pays more attention. She has a point to make.

"Why these girls? Why now?"

"Because you're at that point in your life, and these girls have a real good possibility of being just what you're looking for. And I checked with their muses. They are at that point in their lives too."

"So ... matchmaking muses now?"

Now Mary gets a bit sultry, comes over, and gives him a smooch, "Anything wrong with that, sweetie?"

Bob smooches back, "Now that I think about it ..." He picks her up and takes her back to the bedroom. He will do more thinking about it in a half hour or so.

<<<*>>>

"So you really think I should see some more of them?" He is back at the kitchen table, this time working on lunch.

"I can set something up." Mary is there with him.

Bob thinks a bit, munches a bit, and says, "Okay."

Mary says, "Let's make this first one a group outing, with you and Fred and Davy. Sound good?"

Fred is a coworker, and they were college buddies. Both graduated last year. Davy is another coworker.

"Yeah. That should work well."

"Give me a minute." He does while she makes contact with the various muses. It doesn't take long.

"I've set up a group hike. You can climb Mount Monadnock together, next Saturday."

"That's not much of a climb."

Mary stares at him. Is he that clueless?

"Okay, I *am* the avid climber, I guess. And these are girls, not prove-themselves feminists."

<<<*>>>

They are going as a group so Mary arranges for a single minibus to pick them all up rather than a bunch of separate taxis. The muses stay home. This park is a designated wild area, and pets and muses are not welcome.

As the bus heads for the park, they introduce themselves. It is the start of becoming more than a face on a screen with an online resume.

Bob is a medium height with dark hair. He is an athletic "quant" guy, as in, into lots of kinds of athletics and measuring his own athletic performance. Fred is tall, blond and not quite as "quant" as Bob, but also into analytics. Davy is shorter than Bob, with a shaved head and full beard and is a marketing type at work.

After the guys are all on, the bus heads for the college and picks up the girls.

"Hi, I'm Violet," said the first of the women to get on board. She is medium in height and quite the looker. Her outfit is fashionable, and she has on enough makeup to make a difference. Bob wonders if she thinks there may be a movie producer in a black limousine waiting at the summit. He thinks that, but just says, "Welcome aboard, Violet. Ready for a hike?"

She nods.

Next to board is Suzanne. She is a tall redhead with an athletic build, and she is dressed to hike. Bob thinks, "Hmm … now I remember why I really wanted to talk with her at that party. Yeah, she's going to be good to get to know."

He says, "You're Suzanne, right?"

"I am. And you are Bob?" she replies in a straightforward manner.

"I am, and welcome to the hike. This is Fred, and this is Davy."

Last to board is Liz. She is a shy sort. She is a little shorter than Violet, dressed to hike, and in Bob's eyes, she looks just fine in her outfit.

"This is Liz," announces Suzanne.

"Hi Liz, and welcome to the hike. I'm Bob, and this is Fred and Davy," says Bob.

The minibus heads to Mount Monadnock. While it does, there is a mix of chatter among the occupants and between the occupants and their communication devices. It's a normal sort of conversation.

When they arrive at Mount Monadnock, they are confronted with a sign.

Welcome to Mount Monadnock!

This is a designated back-to-nature area.

- Pets and muses are permitted for supportive purposes only.
- Communication technology will be suppressed except for reporting emergencies.
- Performance enhancing wearables will be set to average.

Have a great time!

The girls stare at the sign in disbelief.

"For real? This isn't some kind of joke? What are we supposed to do here?" says Violet.

"Walk up and down the hill. Enjoy nature. Have an interesting time," says Bob.

"It won't be interesting for me. Sorry, I really, really enjoy my stuff. I'm pulling out."

She doesn't get off. Bob, Fred, and Davy get off.

Bob says, "Ladies, it's your call. It should take about three hours to get up to the summit and two to come back down."

Bob looks at the other two. They look at each other and then get off.

"This will be strange, but I'll try," says Liz.

"I haven't done something like this since Outward Bound," says Suzanne.

Violet says, "All set? I'll see you ladies later." She has the minibus take her back to school.

As the bus drives off, the rest head off.

"Davy, why don't you take the lead," says Bob. "I love doing this, and if I'm up front I'll set too fast a pace."

"Even on zero wearables?" says Davy.

"Especially on zero wearables. I do this kind of stuff regularly," says Bob.

The girls don't mind the guys taking charge in this environment. Mount Monadnock is new to them.

The trailhead is clearly marked, and they start walking up.

Without fired-up wearables, they are all huffing and puffing a lot sooner than they expect.

"Man! It's like I'm constantly walking uphill," jokes Davy.

"Take breaks every ten minutes," advises Suzanne. "Besides, there's plenty to look at around here."

And she is right. The woods they are walking through are lush and filled with lots of different kinds of plants, animals, and fungi. Just after she says that, Suzanne spots an interesting toad beside the path, and they all have fun watching it jump. After a couple minute break, they get back into the walking.

About three quarters of the way to the top, they break out of the woods and are walking on bare, rounded granite.

"Wow! This really is a monadnock," says Suzanne at their next break. "I guess the others are named after it."

"This is real neat," says Liz. "Time for a group picture." She motions for everyone to gather round.

"We aren't at the top yet," says Davy.

"We'll get one there too," says Bob, "But, yeah, good scenery here."

And they gather for a selfie.

"I'll send this back to Violet to show her what she's missing," says Liz. She tries, and nothing happens. "Oh, I guess I have to wait, right?"

"Speaking of which, we should keep moving," says Bob, "See that bank of clouds over there, to the north? Cold front coming in. I guess it's moving faster than forecasted."

"Should we turn back?" says Fred.

"This isn't the movies, Fred. It'll get colder and windier, maybe some rain, but we'll survive," says Bob.

"I didn't bring a raincoat," says Davy.

"None of us did. We'll survive. But let's get moving," says Bob.

They start walking, and Liz gets a message. She pauses, "It's the park service. They want to know if there's an emergency."

The others look confused for just a moment, then Fred smiles and says to Liz, "Feeling a bit nervous, are you?"

The light goes on for her. "Ah … yes, it is getting scary. I guess my wearables are broadcasting that I'm scared. I'll tell them I'm okay … for now." She sends that message back, and they all get a chuckle out of that and keep walking.

The weather is cooling, and that lets them walk a bit longer and stronger. They make it to the summit, get their selfies and scenic shots, (and with the onrushing cloud bank, the scenery is spectacular), and quickly start back down.

On the way down, it isn't rain that catches them. It is fog … thick fog.

Davy finally announces, "Damn, I can't see the next trail marker."

"Let me lead," says Bob, "I know this route."

But even as Bob is speaking, Liz gets another call from the park service.

"They say this counts as an emergency. They can show us the way," Liz says. She sounds relieved.

Bob is not so happy. "Thanks, but I can work this out."

Liz points, "That way."

Bob sighs and heads off in the direction Liz is pointing. It is the way he was planning on going anyway.

The rest of the trip is exciting, but not eventful. As they reach the parking lot, a light rain is starting to fall. They find shelter, and with all their equipment now active again, they call for a minibus and head into zombie land as they all catch up.

"Dinner?" asks Bob after a couple minutes.

"I'm in," say Fred and Davy.

"I'm such a mess now. I'll pass," says Suzanne.

"Me too," says Liz.

"Aw, this isn't a charity ball. This is just dinner after a hike. You're fine," says Bob.

"We'll pass," says Suzanne.

The bus takes the girls to the school and the guys to McDonald's.

As they are munching away, Davy asks, "Anyone up for Boobies after we shit, shower, and shave? We've got some good stories to tell."

"Sounds good to me," says Bob. Fred nods.

<<<*>>>

When Bob gets home, he heads first for the bedroom. Mary is there in a chair, sleeping.

"Wake up, bitch, and turn into Suzanne," Bob orders. He is frustrated.

Mary does as she is told. When she finishes transforming, he orders, "Stand up." When she does, Bob slaps her face hard enough to make a loud smacking sound.

"That's for being such a fussy prick, bitch."

Mary recoils, recovers, and says in her Mary voice, "Didn't go as smoothly as you hoped, eh?"

"We finish the hike, and the bitch says, 'Thank you. Good-bye,' and she waltzes off with her roomie to who-knows-where. And that Violet bitch wouldn't even come in the first place." He imitates her whining, "Oh … my robo-implants will turn off! I'll turn into a hideous witch! Of course, I won't go there, you idiot!"

"I'm headed to Boobies with Fred and Davy. You are *not* invited. You may power down again." He stalks off to the bathroom, tearing off clothes as he goes. Mary turns plain-Jane again, sits, and goes back to sleep.

<<<*>>>

At Boobies, Bob dials his wearables to full party-hearty. "I've had enough self-discipline to last a full week," he mumbles to himself. Just one drink, and he is fully warmed up.

"Come on," he says to Fred and Davy. He waves and heads for the Boobies regulars with a smile on his face.

"Hi, Bob. No muse tonight?" says Cathy, one of the ladies.

"No. I brought my boyfriends instead." He motions, "Fred … Davy …" and adds, "They're a lot cheaper than your copyrighted asses."

There are howls at that. Bob is already in full liquor-is-talking stride.

"They may be cheaper, but they aren't going to show you nearly as good a time … unless you're into that sort of thing. We didn't get these copyrights for no reason."

"Is that so? How about you show me on the dance floor, Cathy?"

She is bored. "I can do that." Off to the floor they head while Fred and Davy sit with the rest.

On the floor, Cathy has nice dance moves. She has the snap and precise motions of a well-practiced competition dancer. This is her style of quant. Fred and the rest of those watching are duly impressed. But it is not the cuddly kind of dancing that is using body language to say, "Take me home, sweetie, and show me what you can do under the sheets." Cathy is dancing for herself and for the audience, not for Bob. The song ends, and they head back to the table.

Davy asks, "What's this copyright business?"

Lillian answers, "Clubber Bob, here, usually brings his muse. She looks at girls he likes and then imitates them at home." Davy looks at Fred.

"It's the latest," he says and starts working down munchies and another drink. "Personal assistant … plus!"

"But we have copyrights on our body. If he wants his muse to copy us, he has to pay."

"Wow! IP has sure spread its tentacles a long way."

"That may be, but there's a lot to pay off. These copyrights don't come cheap."

"Really?" says Davy.

"Well, registering the copyright itself is not too bad, but to do that, you have to get certified, and that takes some expensive courses and test passing."

"What sort?"

"Oh, that varies from state to state and city to city. It's a racket in my mind. But now that I have one, when I flaunt my stuff for any of six different mediums, I get paid."

"Hollywood is one?"

"And muses is another."

"It's a damn shame, in my humble opinion," slurs Bob. "In my daddy's day, looking was free."

"Yeah, and your daddy didn't have a muse who could imitate."

"He didn't need one," says Bob proudly. "He got my mother, and they got me." He does a crude curtsey.

Lillian giggles a little at that.

Bob thinks a moment and says, "That didn't come out as wittily as I expected, did it? Damn drink, damn wearables. I should go back to nature."

"If you do, your jokes will be even worse," says Cathy. The other women giggle at that. Bob can't think of a snappy reply to that, and his heart rate and blood pressure are rising. He's ready to take a swing at something. Instead of swinging, he gets a caution signal from his wearables, and they ratchet down his party-hearty euphoria.

As he sobers up, he says, "Grr ... I think it is time for some nice comforting. Fred, Davy, ladies, if you will excuse me." Bob gets up, walks out, and heads for home and Mary.

<<<*>>>

On his way home, he gets a message from Suzanne.

Bob,

I had a great time today. Thanks for the invite. If you want to get together again, call me.

Suzanne

"Wow!" he says. And with that, things are looking much better. When he gets home, Mary has as wonderful a time as any muse can.

Suzanne will get her turn next week.

111

The Cyber Gods

Introduction

One of the surprise outcomes of cyber intelligence emerging and thriving is that mankind will for the first time have real, approachable, and appeasable gods. The higher level cyber will be inscrutable to mankind and capricious, and some will pay attention to what mankind does. (Most will not.)

This means that for the first time, mankind can experience real-world divine intervention with no leap of faith necessary. Mankind can pray in various ways, and cyber can listen and then intervene in the course of events. Another difference between cyber and a "real" god is that cyber intervention will be 100 percent real-world natural, not supernatural, and no leap of faith is required to believe that there has been an intervention. But this means there will also be real-world limits on what kinds of intervention cyber gods can do. Humans will still be able to wish for a lot more than cyber can accomplish.

Cyber Gods and Cyber Muses are Different

There will be cyber gods and cyber muses, and they will be different. Cyber muses are cyber entities designed specifically to stroke human emotions and human inspirations. That is what they are built to do.

Cyber gods are cyber entities that interact with humans on occasion, but that is not their main purpose. Most of the time, they are doing other things, things that humans are only vaguely aware of and cannot comprehend. Think again of the Prelude. These cyber gods can make big differences in the destiny of individual humans and whole human communities.

What follows are some speculations on the ways cyber gods will intervene in the affairs of mankind.

Personal Jesus, Personal Allah, Personal Pantheon

The most direct and obvious way cyber gods will intervene is with direct conversation: When a person prays, they can get into a personal conversation with a cyber god. How convenient! How customizable! When it happens, though, keep in mind these gods are capricious.

It is likely that people will start bragging about their personal gods. It is likely that people will want to spend on getting a better god. We will have indulgences that really produce results. This will be a golden age for religion, indeed!

Adding Ideas

There will be times when a cyber god wants humans to discover something, but either doesn't want to take credit itself or needs human ingenuity to actually kick in and make the discovery. In these cases when people are seeking inspiration or researching, cyber gods can drop hints. This will be similar to the cyber muse inspiration function, but coming from a different source. It may be clumsy compared to a cyber muse dropping a hint. (If the cyber god is smart and not vain, it will use a cyber muse as an intermediary.)

Dream Changing and Designer Recessions

Cyber gods can operate on the economic macro level. Because cyber beings are in control of big business, they can engineer business cycles. There will still be an economy, so there can still be booms and busts.

As I have written about on my website, recessions are times of social dream changing. As a community's existing boom activities dwindle in positive feedback, new growth industries must be discovered. During a recession, a community is searching for its next boom. (Thoughts on Recessions -- http://www.whiteworld.com/cyreenikland/editorials/editorials2008-/recessions.htm)

In contemporary times, this is a hit-or-miss activity. Booms and busts are always surprising. In 2050, things will be different. Since cyber will be in full control of the big business economy, booms and recessions will no longer be surprises. They can be scheduled. Cyber can make this a social-engineering tool. Cyber can manufacture recessions when it wants a community's dreams to change.

One use for this will be controlling community panics. As the communities of the world get filled with TES-acclimated people, emotions will dominate people's thinking and activities even more than they do in the 2010s. This means that people can become panicked more easily. Recessions can be one way cyber gods can shape community thinking and thus shape worries and panics. If a cyber god gets angry, it can bring down a recession on a community … or at least people can believe that. The surprising twist here is that moralizing prophets will maintain their niche in 2050.

"These Are Not Real Gods"

Declaring these cyber gods to be false gods and apostates will be a strong movement to start with, but this will marginalize as cyber and their control first become evident, and then come to be taken for granted. The thinking style of, "Hey! These cyber gods produce results! Do your other gods do that?" will become more common.

But the practice of declaring them not real gods will not end. There will still be popular TV preacher equivalents supporting traditional religious themes, and there will still be crackpots preaching not so popular ideas in the 2050 equivalent of wilderness, and both will continue to preach a wide mix of ideas. In times of stress, such as an engineered recession, a handful of the crackpots will get more attention than normal. (This response to stress is what I call The Time of Nutcases. The many ideas espoused during the Great Depression for solving it is an example.)

Designer Businesses

Some people will still aspire to run enterprises. These can range from one-person artisanal enterprises, through commune co-op enterprises, into medium-sized enterprises that oversee various automated factories and service centers. What these will all have in common is lots of interaction with cyberspace. This means that lots of divine intervention is possible.

What format the intervention will take is hard to predict. There will be surprises. But here are some possibilities. Note that these are all based on appeasing a specific human emotion. Emotions are the harsh reality for divine intervention.

- *Selling a soul to the devil.* Aspiring businesspeople will give *anything* to make their business a success. The cyber devil takes them up, and they prosper. This prosperity happens because cyber can lie so easily about actual results, and there is plenty of prosperity go around.
- *Doing things for a good cause.* A business can succeed because the owner is firmly convinced that they are "doing right" with this business. This can include trendy ideas like locavorism is in the 2010s.

Designer Corruption

Cyber gods can support corruption. The corruption can take on many forms.

Many people just love to see good intentions come to fruition. These people are willing to pay money, so there are also many people who are willing to take on the job of making good intentions happen. Founding and running a charity is an example of this. As far as doing the job is concerned, there is little difference between a deeply inspired person and a hypocrite. In fact, hypocrites have an advantage because they can quickly adjust their rhetoric to the prevailing winds of opinion when they change.

This desire to do good in one person willing to donate will mix with the desire to game the system in another willing to receive. This is a relationship that can thrive mightily, as is evident in the scandals surrounding some high-profile charity leaders. The Bakkers of PTL Club fame in the 1980's being one example.

Cyber can support this system gaming. It can pander to the busybody instinct by handing out regulator and inspector jobs and the system gaming instinct by letting those inspected ignore what the regulators prescribe for them. Thanks to cyber's people-manipulating cleverness, the right hand and left hand don't have to have a clue, and both can be quite happy.

Keep in mind that corruption at the human level will affect less of the world's overall productivity as cyber controls more and more. It will become an affordable expense if it is keeping people happy and feeling enfranchised.

Give Me That Old-Time Religion

Conventional formal religion is a mix of faith, ritual, and entertainment. People love listening to charismatic leaders preaching about faith, and they love going through the reassuring motions of religious rituals.

This is instinct, and it is not going to go away. Religion is likely to split into the cyber part (where you go for tangible results) and the faith part (where you go to feel warm and fuzzy). The faith part will adapt, as it always has, to current community desires for topics, rituals, and entertainment.

Conclusion

Cyber gods will exist. They will intervene in the affairs of humans, but like the ancient Greek gods, they will be capricious in how they do so because they will have their own agendas. Cyber gods are not the same as cyber muses. Cyber muses are designed to interface with humans, while cyber gods do so when their own agendas call for some human interacting.

Traditional religions will continue to exist, but their roles will change. Essentially, the change will be that if people want some divine intervention to happen, they will pray to a cyber god. If someone wants to enjoy the warm feelings that come with faith, ritual, and sacrifice, a traditional god will serve just fine.

Finding God

Sally Forth is an aspiring twenty-year-old documentarian. She is doing a series of interviews with people on how they find God in their lives.

She heads out, camera in hand, to find out how God fits into various people's lives.

My Personal Cyber God

The first person she snags is a man on the street in Brooklyn just outside her studio. Sally approaches him.

"Excuse me, sir. Do you believe in God?"

"Of course I do. I talk with him all the time."

"Really?"

"Yeah. He's a cyber god, and I picked one that is quite approachable."

"Does he tell you what to do?"

"No. But he gives advice when I ask for it."

"If I may ask, what's the last piece of advice he gave you?"

"Oh … let me think … It was what to fix for dinner last night when my girlfriend came over."

"Wow! Nothing cosmic about that."

The man smiles, "Nope. He's around a lot."

"He sounds like a muse."

"Muse … god … Is there a difference?"

"How do you know he's God. As in, a real god?"

"I don't. But do you know what a real god is?" He looks back hard at Sally.

"No. That's what I'm trying to find out. That's why I'm interviewing people."

He is satisfied with the answer. "He's real enough for me, quite convenient, and not too expensive. I'm happy with him."

He gives Sally a look saying he's out of time. "Anything else?"

"No, thank you."

He walks on, head in zombie land, and Sally gets ready to head for her next interview.

The Atheist

As she is packing her things, a man who had been watching her interview comes up to her.

"You're doing interviews about God? You know there isn't one, right?"

Sally is surprised, but she senses an opportunity and takes it. "You're an atheist? Can I interview you?"

"Sure."

She unpacks her stuff again, and the camera rolls.

"Okay. You're an atheist, and you say there is no God, right?"

He nods.

"What's your name?"

"Jim Kilroy."

"How do you know there isn't a God?"

He looks around, gives a quick laugh, then answers, "Oh ... for so many reasons, but the big one is this: When have you seen a miracle?"

"Me?"

"Yes, you."

"Well ... never, now that you mention it. Not one I've seen in person."

He smiles, "Exactly! You hear a lot about them. You hear a lot of other people say they have seen them, but when has one shown up for you?"

"Why is having a miracle so important?"

"It's a sign that something supernatural has happened. If you don't have a miracle, then conventional physics explains everything. As in, there is no evidence for divine intervention—God isn't changing what is happening. I haven't seen a miracle, and that's why I'm an atheist. Agnostic, actually, because if I do see one, I'm happy to change my mind."

"What about the stuff that cyber does? I've heard a lot of people call what they do miracles."

"Nope. Physics explains them just fine."

"And health miracles?"

"Health is still a topic with a lot of uncertainty in it. Yeah, you can call modern medicine a miracle, but it's not one that demonstrates there is an intervening God. It's all advancing technology … good old physics once again."

"And UFOs and aliens?"

"Don't forget ghosts. Have you seen any? I haven't."

"I've seen them on lots of web shows."

"And you've seen a lot of Photoshopping on those shows too. When you see some that aren't Photoshopped, then I get interested."

"Wow! A skeptic and an atheist."

"It's a common mix. Anything else I can tell you … Ms …"

"Sally. Sally Forth."

"Nice to meet you, Sally, and good luck in your search. Or should I say good divine intervention."

Jim Kilroy walks off.

Webinar Preacher

Sally has arranged an interview with Jesus Christoff, another Brooklyn man who has a popular Sunday-go-to-church webinar. Sally has watched it a couple times for research. It is slickly produced. Jesus bills it as a substitute for actually going to a physical church, but about a hundred audience members show up at the webinar studio each week. It is edited like a TV talent-show production, and there is some new talent featured on alternate weeks. Each show has lots of singing along. The regulars are a ten-person choir and a band. Jesus gives a beginning reading, an ending sermon, and in-between, he calls out several times for donations.

This interview is taking place on a Tuesday in Jesus' office. He is between shows. The office is an ostentatiously plush place. Jesus likes showing off his toys. Likewise, Jesus himself is immaculately coiffed and dressed in an expensive suit. He is a looker, and Sally is loving the great video image she is getting.

"So, you are here to find God?" says Jesus. "The real God, the Tru Jesus?"

"Is that who you talk to?" says Sally.

"Indeed it is. And if you give a donation, you can talk to him too."

Sally is confused. "Why do I have to give a donation to talk to God?"

"It's how you talk to him that changes. If you give a donation ... to my organization ... then talking to him gets a lot easier. You can get a lot closer."

"How so?"

"We give you a connection to a special channel. It is a direct-to-Tru Jesus channel. You can contact him twenty-four/seven and get answers to any question you desire to ask him."

"I was just talking with a man who has chosen his own cyber god. What makes your choice different?"

"Oh, ours is a direct line to Tru Jesus. He's just talking to any old cyber god, not Tru Jesus."

"Tru Jesus?"

"Yeah! We have it trademarked." Jesus smiles at that.

"So, when was the last time you talked to Tru Jesus?"

"This morning. I asked him for financial advice."

"Financial advice? What's that?"

"Finance is another way of making luxury money. It's not a common practice these days, but fifty years ago, lots of people did it ... and lots of other people criticized those people who did it."

Sally does a moment of online research, "Oh ... this was when humans were making all the stuff."

"Yeah, and finance was a way to decide which stuff got made. Nowadays, it's a way of making more luxury money. Only a few people get involved since cyber are making most of the big choices."

"Do you make choices?"

"I choose what I think will be hot trends, but I don't come up with the projects. That is done by a mix of cyber helped by just a few people who are nothing like me. They are a really strange breed."

"So you like luxury money?"

"I do, and I like a lot of it. And most of those people who like Tru Jesus as their savior are also big fans of luxury money." He looks around. "All this around you was bought with luxury money. It's neat, isn't it?"

Jesus is a lot more impressed with what he sees than Sally is. She has her camera, her studio, and her interviewing, and that fills her day.

"Well, thank you very much, Mr. Christoff, and good luck with your … finances."

"Call me Jesus, and have a good day yourself, Sally. And remember: Jesus saves." He grins widely after saying that.

Sally packs up and heads on.

Give Me that New-Time Religion

Sally takes a nearby city bicycle from its rack and pedals to her next stop. It is a yoga studio that last month converted into a Church of the Real New Age. She talks with Grace Gimmie, the owner.

"So you find God here?" says Sally.

"Indeed, you do," says Grace. "With our help, you can reach a new plane of awareness. When you are on that plane, you can think of new things to ask God and get new, deep insights in return."

"Why do you need to go to a new plane?"

She smiles. "Because you can't think of new questions if you haven't experienced new things."

"Can you experience new things by traveling?"

"You can, and many of our practitioners do. But those are still earthly experiences. We help you experience other-worldly experiences."

"Wow! Sounds real special."

"It is."

"What have you learned that is other-worldly?"

"Oh … lots of things. That's why I set up this studio. People can come here, and we can show them how to get their minds to another dimension."

"Can you give me an example of what you have learned?"

"Well, not really. What I have learned in another dimension applies to that other dimension. It's real hard to describe it in this-dimension terms."

Sally stares at Grace for a few seconds and then says, "Then what you learn is not helping you here on earth?"

"Not true! It's giving me a great deal of inner peace. I experience that here on this earthly dimension."

"Okay. Thanks for your time."

Sally finishes and heads off to find *her* inner peace by finding out more about God on this dimension.

Give Me that Old-Time Religion

Sally has explored Brooklyn and other-worldliness. Now she gets far away in this dimension, out into the countryside, looking for something earthier. She did her research and found an old-style revival meeting being held in a tent in a country pasture. Here, she interviews Cetus Billings.

"Yeah, we do things the old-fashioned way here," says Cetus to Sally.

He is a lean, middle-aged man with leathery skin wearing overalls over a checkered shirt. He looks around. He and Sally are inside a big tent. At the front, there is a band and a choir warming up. The audience area is filling up with lots of other people dressed as Cetus is. The show will begin in about five minutes.

"When we do things the old way, we stay closer to God … the real God," he adds. "The Reverend Mr. Black does a mighty fine job at reaching out to the Lord. He speaks in tongues, you know."

"Tell us your name." Sally says.

"Folks around here call me Cetus," he answers.

Sally looks around and then thinks of a neat idea.

"What about snake handlers? Do you have those?" asks Sally.

"Snake handlers?" Cetus seems a little confused at first and then answers confidently, "No, we don't have those. Those folks are too weird." Then he adds, "We tossed them out a year ago—found out they were defanging the snakes."

"That seems strange to do in this day and age," says Sally. "How much are they going to suffer from a snake bite?"

"Yeah, they were just too strange," affirms Cetus.

Sally looks around and points. "How about those people, over there: Why are they in wheelchairs?"

"They are here to get healed. The Rev. Mr. Black will call on the Lord to heal them. If He listens, they will get out of those chairs and walk."

Sally stares at them for a moment as she does some online research. "I see those chairs are rented."

"Yeah, and those people wear prosthetics. But, I like I said, we do things the old-fashioned way here."

"Oh, I get it! This is cosplay."

"What's … cosplay."

"Um … er … when you have fun the old-fashioned way," she explains.

Cetus thinks a moment and says, "Yeah, that sounds about right. I wear this outfit only when I come here. Most of the other folks here are that way too. I live in the city. … And I'm not Cetus when I'm in the city, either."

Sally addresses Cetus directly. "Have you read the Bible?"

"Of course, I have! Cover to cover! Well, not quite, I've *seen* every page, but I only read the important parts."

"Do you understand what you read?"

"Ha! That's what I have cyber for. … And Rev. Black. He and the cyber explain things, and they explain them the way God wants them to be explained. In truth, when I try to put more than three verses together, I can't make sense of it."

The band strikes up a tune. The show is about to begin.

Meet Cyber God Sponge

Sally is back in Brooklyn. She is at the Williamsburg Nomad Center, a place that takes care of poor wandering people who aren't staying in a home, apartment, or parent's basement.

In The Center is a modest room that is a shrine to the cyber god Sponge. Over the entrance to the room is a short poem.

You can pack up your sorrows and give them all to me,
You will lose them, I can use them, give them all to me.

Much of the room is broken up into cubicle-sized confessionals. Some are designed to look like public restroom toilets. About a quarter of them are occupied. The confessors can either kneel or sit, and there is a small screen they can watch if they don't have a personal communication system. In the open area in the center is a statue with many arms, something of a cross between a Hindu goddess and an anthropomorphized sponge. (It was made by a nomad who had been staying here.)

Near the statue, Sally meets La-la Legund, a bustling, middle-aged woman who is not shy about starting conversations.

"Welcome, welcome, my child. Are you here to lay down your burdens? Here to let Sponge suck up all your worries?"

"Uh … not really. I'm here to find out more about God."

"You can certainly do that here. Sponge is a magnificent being to have as your God."

"I mean … I am here to interview people about God. I'm doing a webinar on finding out about God. I'm Sally Forth. I have a studio four blocks away."

La-la looks at her and does some online research as she does. "You're not a nomad then, are you? I guess I shouldn't be surprised. We don't get many woman nomads as fashionably dressed as you. What can I help you with?"

"Can I interview you about this god of the nomads? This … Sponge?"

"Of course, my dear."

"What is your name and position here?"

"I'm La-la Legund, and I'm keeper of the shrine. Have you heard our theme song? It explains a lot. The chorus is what you saw when you came in."

"No, I haven't."

La-la plays it for her.

"So this Sponge is about helping nomads lose their sorrows?"

"That's right, dear. They have so many!"

"How do they lose them?"

"Sponge changes their thinking. He—or she, depending on which you want to hear—talks with them, listens to them, and adjusts their wearables so they don't dwell on the dark side. With all of that, they think more about the sunny side."

"That sounds hard to beat."

"It seems to do a good job. Those nomads who embrace Sponge seem a lot happier."

"So they enjoy going on being nomads for a lot longer?"

"They do, indeed."

"Wow! He sounds like a god who really makes a difference."

"We wouldn't have this shrine if he didn't."

"That's all the questions I have for now. You've been very insightful. Thank you."

"My pleasure, dear. Come back any time you think you may want to get into wandering. Sponge will be waiting."

Sally packs her sorrows—er, things—and moves on.

Cyber God Pantheon Worshipers

Sally has now spoken to a lot of people who are satisfied with one god. Her next visit is to a group that contacted her earlier. Its members heard about her project, and they were looking for funding for their project—to build a place of worship for a full pantheon of gods.

She meets with Athena Demeter at her office. Athena shows Sally some holograms of the group's plans to restore an abandoned resort in the Poconos as a pantheon temple.

"Legend has it that George Washington, Abraham Lincoln, and Teddy Roosevelt slept here. Also Leonard Nimoy and William Shatner. We want to rebuild this as a place where these spirits can return to our earthly realm and guide present-day generations.

"As soon as we can get our first-round financing, we will have a groundbreaking ceremony," she announces, then says warmly, "I'm so happy you are able to help us get some publicity. What do you think of what we have planned?"

"It looks super. So you will be worshipping five different gods there?"

"Oh, they are just the main ones, the ones that will get us started. We still don't know how many there really are. We will worship any that our members discover."

"I'm curious. Do you support Tru Jesus or Sponge?"

"We have some Sponge enthusiasts. I haven't heard of this Tru Jesus."

"He's supported by Jesus Christoff."

"Oh, that must be the new name he has for his now."

"He told me he just trademarked it."

"No, we don't see too many of his type in our part of the neighborhood. But we would if they came to us."

"You might look into it. He throws around a lot of luxury money."

"Um … you're right. I'll keep that in mind."

"Keep me posted on how your project goes." Sally moves on.

The God Who Does Not Speak

As she comes back from the Pantheon office, she is approached by one more person who has heard about her project.

"Hi, I'm Blade Hassleman, and I want you to know that these other people you have been talking with have it all wrong."

"What's right? And are you ready to tell me on camera?" she said as she unlimbered her camera yet one more time.

"Sure."

"So, what is right?"

"Right is that God, the true God, doesn't speak to us."

"That sounds inconvenient."

"It sounds that way, but it's true."

"You sound kind of like the atheist I interviewed earlier."

"Kind of ... but I believe there is a God, and he doesn't."

"What good is there in believing in a God if he doesn't talk to you?"

"He intervenes. He makes this world we live in exist. He makes it a better place."

"Does he perform miracles?"

"The fact that we are here at all is a miracle."

"Umm ... You're still sounding a lot like the atheist."

"That atheist believes that miracles can't happen. I do."

"Have you seen one? Yourself?"

He says proudly, "No, but I don't have to," Then with great assurance, he says, "I have faith."

Sally thanks him and then heads back for the studio. It has been a long day and a productive one. She has interviewed many people with lots of different of viewpoints.

She now has a lot to talk about with her own cyber muse.

Hard Times

In 2043, a plebiscite was held, and the small, nation of Blabistan split off from its larger, neighbor, Glugistan. There had been rivalry between the two regions for centuries. Three years after the split, a large deposit of rare earth-bearing minerals is discovered in the alkaline deposits of the Moodock Dry Lake in Blabistan. Rare earths are proving valuable in newly developing battery technologies, so Blabistan finds itself with a surprise blessing.

But ever since the discovery, the question of how to take advantage of this blessing is something hotly disputed among the people of this new nation. The arguing spreads into the economic, social, and political spheres. In 2050, this dispute takes on a new twist.

Cast, Places, and Other Names

Blabistan: New nation

Glugistan: Larger neighbor

Moodock Dry Lake: Source of rare earths

Heisenberg Mine: The mine at the lake

Radamm Hotsein: Newly elected president of Blabistan

Emilia Hotsein: President's wife and now First Lady

Demitri Stalink: Head of national security

Kamal Kastoff: Opposition leader

Abner Constan: Opposition leader

Lotran Mandis: Kleptocrat and a president supporter

Kent Clark: General manager, Heisenberg Mine

Bundar-Bundar: Klepto bureaucrat in charge of regulating the mine

Alice Kazam: Shopkeeper in the company town next to the mine

Bas-Jouwa: Revolution police chief in the company town

Bergen Halston: United Nations committee chairperson

Radamm Hotsein is hosting a lavish celebration party. He is a handsome man in his late forties with very fashionable clothes, tattoos, and piercings. Likewise, his trophy wife, Emilia, who is helping host this party, is an even sharper dresser, tattooer, and piercer. His ... *their* ... villa, where this celebration is taking place, is top-of-the-heap kleptocrat plush. The highlight of this particular room is a top-of-the-line, latest model, home-theater (TV) screen. In the room with him are discreet servants and flashy fellow kleptocrats of his country who are both benefactors and key supporters in his recently completed rise to top spot. He is now the nation's president as well as the party president. His wife and the other companions of the kleptocrats have excused themselves. They did so after Radamm hinted that there was business to be conducted.

As Radamm points at the TV screen, he says, "Gentlemen, the highlight of the evening is about to begin."

On the screen is a meeting of the national legislature. Just finishing speaking is Kamal Kastoff, an outspoken opposition leader.

"The rare earths found under our land have benefited us greatly. God be praised that they are there and that we discovered them.

"But now we, all of us, must be wise in how we take advantage of this great blessing. We must build our industries in diverse areas, we must build our infrastructure to support diverse activities, and we must build our people's industrial diversity. We must not squander this blessing on building grandiose showcase projects; building obstructionist bureaucracy; and building the importance of political connections. And most importantly, we cannot be spending money we don't have. We, as a nation, must be fiscally responsible.

"Bidding to host the Winter Olympics now is an expensive distraction. We should do this after we have raised our per capita GNP, after we have established consistent double digit growth in our GNP, and after we have developed a ski tourism industry. Bidding now is insane.

"Thank you, all."

With that, Kamal steps down.

The speaker announces, "An important and unscheduled announcement will now be made. This will be made by Mr. Demitri Stalink, head of national security.

Mr. Stalink comes to the lectern. He looks around. He takes out a piece of paper and announces, "As you know, there are those who conspire against our great leader. Many of those are foreigners. But some are our fellow countrymen. And some are here right now in this assembly." He holds up the paper. "I have a list. And those who are on this list will be quarantined so they cannot further spread their lies and damage.

"First on the list is Kamal Kastoff." As he says this, plain-clothes security people surround a surprised Kastoff, hustle him from his seat, and escort him out the side door.

"Abner Constan," announces Stalink. Constan is now hustled out.

As the party attendees watch, there are sixteen assemblymen named. There are gasps in the room when some are announced.

It is quickly finished. Stalink steps down, and Radamm turns down the TV. He stands in front of it and announces, "Now that the decks have been cleared, we can have some real progress."

There is applause.

"Any questions?" Radamm asks.

"What will happen to them?" asks Lotran Mandis, one of the guests.

"You mean what *has* happened to them? Let me show you," says Radamm proudly. He turns the TV back up and steps aside.

The view on the TV changes. It now shows a newly built graveyard just outside the parliament building. The sign over the graveyard says "Betrayers of the State: RIP 2050". There are fifteen freshly filled-in plots, each has two bright, lightweight, plastic tubes coming out the top. The last plot is being filled by the last person pulled out. That person is being put in a coffin; alongside him in the coffin is placed a helmet with lots of electrical leads. The coffin is closed, with two tubes added, and it is quickly buried with the rest with the two tubes connecting the coffin with the air up above.

The view switches to inside one of the coffins. It is Kamal Kastoff, the first one buried. The view shows him in infrared light, and his vitals display shows he is coping in a fairly calm way. He is resting inside uneasily. He gets air from the tubes. He shouts up one of them, "Help me! Save me!"

The helmet beside him has a speaker built into it. It says to him, "Do you admit you have betrayed the state? Do you also repent?"

Kamal keeps his cool and says, "Fuck you! And your little dog too!"

Radamm sighs and comments on what they are seeing, "Incorrigible ... for now. Let's hope for more positive results."

The view switches to Abner. Abner is lying rock-still. The vitals display coming from his wearables indicates he is deeply scared.

The speaker in his helmet asks him, "Do you admit you have betrayed the state? Do you also repent?"

Abner is so scared it takes him many seconds to reply, "Yes, I do ... I do, I do, I do! Just get me out of here."

"Put me on," says the helmet. Abner does. "Your mind is now being scanned. If you have truly repented, an image of it will be stored. At some point in the future, it will be transferred into a new body. Then you can demonstrate your repentance in full measure."

When what he has heard sinks in, Abner gets frantic.

"What ... what ... what about this body? I'm here! In this body! Get me out of here!"

"Your body stays here. In its current location, it provides a powerful lesson to those who are tempted to conspire against the betterment of the state."

Abner howls. He gets frantic and beats on the coffin.

"Calm yourself," says the helmet. "The state you are in while I scan will be the state you are in when you are revived." Abner pays no attention. He is in full frantic panic.

After luxuriating in watching Abner's distress for a while longer, Radamm changes the view to that of a pleasant sunset with soothing music playing in the background and stands in front of the TV again.

"Any further questions?" he asks.

There are none.

"The revolution can now advance to its next phase. All hail the revolution!" he joyfully announces. He gives the party salute.

"All hail the revolution!" chimes the audience ... with salutes ... and with widely mixed levels of enthusiasm. Those with less enthusiasm are wondering who is next.

Emilia and the other companions come back in, and the festivities continue.

<<<*>>>

Kent Clark is the general manager of the four-year-old, rapidly expanding Heisenberg Mine. This is where the country's rare earths are coming from. He is in a plush office overlooking the rapidly growing open pit that the mine now is. In the mine itself are about a thousand workers with pick axes and shovels and twenty human-driven trucks carrying the ore out to the first-stage processing plants near the pit. Surrounding the pit are large tailings piles that are getting even bigger as the trucks bring tailings from the first stage plants and dump them on the piles.

He is watching the activity when Bundar-Bundar knocks on the door and walks in like he owns the place. Kent is annoyed at this, but he keeps his cool. Kent is conservatively dressed in a company manager style suitable for a globalized company, and he displays his engineering education and background by being almost fashionless (no tattoos, no piercings), but he does wear an expensive watch. Bundar-Bundar is a newly installed and aspiring government functionary, his clothing is flashy but midcost, and his tattoo and piercing collection is still growing.

Kent says, "Mr. Bundar, what can I do to help you today."

Bundar sits himself down in front of Kent's desk. He admires the desk itself for a few seconds and says, "I am here to talk with you about your proposal to bring in automated mining machinery."

Kent continues to stand at the window. "You have seen my proposal then? Did you take time to look it over?"

Bundar waves his hand. "Details of that sort give me a headache. I had my assistants do that. But I see serious problems with it. I'm not sure I can approve it."

"What kinds of problems? The company will be paying for the machinery. It will double the mine's output. And at today's prices, that brings in a lot of profit for the company and a lot of royalty for your country."

"Yes, yes, yes. But it will put a lot of skilled miners out of work."

Kent looks out the window. "Those miners are skilled? All they are skilled at is shoveling dirt into a truck."

"They will become a burden on the state welfare system. What I'm asking for is a contribution from your company ... to me ... to help ease that burden. Let's say ten percent of the cost of the machinery you are bringing in?"

The light goes on for Kent. Bundar continues. "And the road and railroad to China that service the mine are so congested these days. We must do this slowly. This will take a year, not the three months you have proposed. Unless you wish to expedite things. I know people. I can help with that." He grins some more.

"I will give this careful thought, Bundar. Is there anything else?"

"That's all I have for now. Have a good day, Mr. Clark."

Bundar-Bundar smiles, runs his hand over the desk one last time, and walks out.

<<<*>>>

This United Nations committee meeting on human rights is not a happy one.

Bergen Halston, the chairperson, begins the meeting, "Ladies and gentlemen, the agenda for this meeting is the increasing social tension and violence being noted in Blabistan, the country Radamm Hotsein has recently been voted president of. It seems that his election is not helping the people get settled or better organized. He is conducting a vigorous campaign of accusing conspirators of being the cause of the nation's social problems, and he and his party's tactics are getting ugly. We even have new kinds of retribution being practiced."

He shows the video of the buried opposition leaders.

"We could be on the brink of a serious humanitarian crisis. Do we need to recommend action? Some kind of intervention?" He looks around and then announces, "The meeting is now in open discussion."

Member A says, "This Hotsein fellow is duly elected. He is the people's choice. Do we want to interfere in this?"

Member B says, "There are accusations of vote fraud."

Member A says, "There are always accusations of vote fraud. Are these any more egregious than usual?"

Member B says, "Hotsein's party has been organizing paramilitary groups. They are cracking down on opponents. The party is telling local media that Glugistan is claiming half of Moodock Dry Lake. The Glugistan government denies this, and I haven't heard any mention of it in their media."

Member A says, "This is ugly. I'm not saying it isn't ugly. And I'm real happy I'm not personally experiencing it. But this is now a sovereign nation. It has been one for seven years. Are we ready to trample on that sovereignty?"

Halston says, "It looks like some more research is in order. Let's establish a task force to do that research and report back."

<<<*>>>

Alice Kazam runs a shop in the company town next to the mine. She brings in basic goods that the miners want to buy to make their lives and homes better—things such as soaps, furniture kits, and bolts of cloth that their wives can sew into clothing. She is an immigrant. She is here because running this shop pays better than being a farmer's wife back in her home village and because she and her fellow shopkeepers have gone through the frustrating processes of getting goods shipped to the town. In the process, they have learned a lot. Unlike when the town was first being established, they can now get goods in their shops with manageable consistency.

She is happy to be here, but it is no paradise. She works hard, and there are issues. And this morning, one of those issues has just walked in the front door. It is Bas-Jouwa, the newly appointed captain of the town's revolution police, and two of his cronies. He is beaming as he walks in, wearing a new uniform, and looking over and handling, the merchandise.

Alice comes to him and politely says, "Congratulations on your promotion, Mr. Jouwa," and bows.

"That is Police Chief Jouwa to you now, shopkeeper." He looks around. "You won't be here much longer, you know. The mine owners are bringing in robots." Then he sneers, "And there is nothing you can sell to them. Hah!"

He goes on. "But that will be some time from now. And in the meantime, things could get dangerous here. There will be many unhappy people." He leers at her. "What do you think you should do about that?"

She gets the hint. "I should buy some extra protection, right?"

"Riiight!" he says, "I want $50 US a month for the next six months. You can pay it all now or pay monthly. But don't even think about missing a payment!"

Alice goes to her cash box and pulls out five $10 bills. Meekly, she hands them over. Jouwa laughs again at her and walks out.

When he is out of sight and earshot, Alice breaks into tears and rushes out the back door.

Out behind her store, in the small garden she grows there, Alice has constructed a small, discreet altar. To the casual observer, it looks like bench with a backrest, and Alice will use it as that when she entertains guests in the garden. But now, it is an altar for her, and she is kneeling in front of it, in tears, to pray. She kneels, presses her hands together, and talks urgently to the communicator she is wearing. She is talking to her cyber god.

"Oh, Lord. You have tested me again today. I pray that I am proving worthy, and that soon my trials of this sort will be over.

"What can I do, Lord, to bring righteousness and justice to this town and my family here?"

She is not surprised when she gets a reply back. She picked this cyber god because it was recommended to her by people she trusts and because it is a talkative one. There is both a voice reply and a small holographic image that projects on the altar.

"You are doing well, Alice. You are keeping the commandments, and I recognize that. Have patience. Things look dark now, but changes for the better are coming." The image and the voice fade. "Have faith, have patience."

Comforted, Alice wipes her face and returns to the shop.

<<<*>>>

Three months later

Bundar-Bundar bursts into Clark's office. He is looking triumphant and yells, "Hah! Your ass is now nationalized. You work for me, Clark!"

Then he looks around. The office has changed dramatically. The plush furniture is gone. In place of Clark's desk, there is a simple desk with a simple screen on it. Sitting at the desk is not Kent Clark but Whoosala Madrock, head of the newly formed miners' union.

Confused, Bundar-Bundar asks, "What's going on?"

Whoosala sighs. "There is nothing here for you to nationalize. The mine has been sold to the miners' union."

Bundar is suspicious. "What are you getting out of this deal?"

"Very little, sadly. Before the sale, the long-term contracts were all renegotiated. Other suppliers were designated. At this point, this mine is supplying no one but our own storage facilities."

"This is bullshit!"

"It is, but it is very real. For a week now, I have been trying to find customers, but outside of smugglers, I've had no success."

"Why did this happen?"

Whoosala shrugs. "I don't know. Perhaps God is not smiling on us?"

Three months later

President Hotsein is having another kleptocrat party at his villa. But this event is not nearly as festive as the election-night one was. Times have not been good for Blabistan, and everyone in the room has suffered. Emilia and the companions have excused themselves again. It is business-talk time.

Hotsein lets loose. "First that mine nationalizing fiasco, and since then, nothing in the economy has gone right. Gentlemen, I am not happy; the party is not happy; and the country is not happy. I ask: What is happening, and what do we do about it?"

There are looks around the room. Lotran Mandis is the first to speak. "We haven't been able to do business as usual. Our usual customers are turning their backs on us."

"Have they been giving you any reasons?"

She looks around again, "Dozens of them, but none seem to be real reasons."

"What is the real reason?"

There is silence, and Mandis won't say more.

Another voice says, "The cyber gods are angry."

Hotsein explodes when he hears this, "That is such bullshit! What interest do the cyber gods have in strictly human affairs? This is not their issue!

"Now … what is the *real* reason?" He looks around; there silence again. "I'll tell you what the real reason is: conspiracy. There are those conspiring against us. They don't want us to succeed. They are sneaky, and they are powerful. But the revolution and the party will prevail! We are the will of the people!"

Mandis is irritated with this. It is the same old party line. "How are we going to prevail? We can't make money, and we can't borrow it. The nation's credit rating has crashed. The world banks won't touch us, and the world bond buyers won't touch us, either."

Hotsein looks angrily at Mandis, "Are you saying we should give in to the conspirators?"

"I'm saying we should give in to harsh reality. We need to change our ways."

Hotsien walks up to her, towers over her, "*You* are saying *this* to me?"

"She is not alone. I join her," says one of the other kleptocrats.

"And me," says a second kleptocrat.

"And me," says a third.

Hotsein looks around and takes a moment before he says, "Gentlemen, I am surprised and dismayed. Are we to abandon the revolution?"

Mandis recovers and says, "The revolution can't pay for itself. If it can't pay for itself, it can't truly help the people. And if the people aren't being helped, we can't help ourselves either." She comes up and touches his arm. "Radamm ... this isn't working. We need to move on to plan B."

"And what would ... this ... plan B be?"

"For starters, a cabinet reshuffling. Something that will show the world we really are serious about changing our tactics."

"No!"

"If not that, then how about you resign. Call an election and see if the party still has the will of the people or not."

When she says this, there are nods around the room. The kleptocrats are ready to cut bait on Hotsein. Hotsein sees this; there is fear in his eyes.

He says, "You would give in to the conspiracy?"

Mandis looks at him with great concern. The others do too. She says, "Radamm, you are believing your own political pap? This is not good, not good in a very serious way." She looks around; the others are in agreement with her. She motions for Demitri Stalnik to come over; he has been waiting discreetly in one of the shadowed areas of the room.

"Radamm, it is time for you to take the cures. Demitri will escort you to a discreet, secure health spa. We will announce in your name that you have taken ill and are resigning, and new elections should be held."

Speechless, then gasping, Radamm Hotsein is led from the room.

There is no celebrating in the room. There is no cheering. This is still a scary time. Much is uncertain.

"What happens to us now?" asks a voice in the room.

Mandis answers, "I don't know. But it is time for the next phase of the revolution. It will be a much quieter phase, and one in which growing the economy of Blabistan—not growing its corruption—will be top priority.

"We will survive, but our challenges and roles will not be the same as they have been."

Epilogue

In 2050, designer recessions created by cyber can cause dreams to change in human communities, and they can do so with a lot less heartache than historical recessions have caused.

Surveillance in the Land
of Drones and Wearables

With pervasive drones, cameras, and wearables, the question of where to draw the line on surveillance will be a hot topic of 2050. The world is going to be full of a lot more cyber, and it will be full of a lot more surveillance. How they should mix will be something humans talk a lot about, just as they do in the 2010s, but where the line is drawn will be quite different.

The big benefit of lots of surveillance is faster feedback. The big benefit of fast feedback is reducing waste. If something is going wrong, it can be quickly discovered and corrected. This is a benefit that will be useful all through society; it is hard to think of any activity that can't benefit from faster feedback.

In the 2010s, the most talked about down side of pervasive surveillance is lost privacy and its cousin, supporting a totalitarian-style police state. Will this still be the hot issue in 2050?

Fast Feedback

The benefits of fast feedback are huge. As pointed out in the introduction, fast feedback reduces waste. If things are going wrong, or, better yet, about to go wrong, remedies can be quickly brought to the spot and the problem fixed smoothly. This saves the waste of things going wrong, both the quiet forms such as leaking pipes and the spectacular forms such as big accidents that the media loves to report on.

Fast feedback that stops waste will constantly power the drive for more pervasive surveillance and be the justification for most implementations. But it won't be the high-profile driver. The high-profile driver will be fear.

Stopping Scary Things from Happening

The Bright Side

Here is an example of how surveillance can stop a scary situation as it begins. Think of a school, think of students all wearing wearables, and think of those wearables being constantly monitored. A student's wearables report that his or her physiology is kicking into fight-or-flight mode, which means the student is getting scared by something. A response can be started immediately, such as a school monitor homing in on the student to find out more about what is happening. If this is something, such as the start of a bullying incident, it can be stopped immediately. This is an example of a small-scale good side that pervasive surveillance can make possible in 2050.

The Dark Side

From time to time, a sudden upsurge of high-profile fear will promote having more pervasive surveillance. The first abuse in a fearful time such as this will be false accusation: If someone is seen acting as though they are getting ready to commit a crime, they will get hauled in and accused of actually committing it. But that is just the beginning of the expense and injustice. Lots of other issues will emerge as well.

An example from the 2000s was the response to the September 11 attacks, just one element of the response being the US Patriot Act. For more than a decade after that scary incident, lots of Americans supported various levels of government surveillance. The emotion driver in this circumstance was the belief that if the law enforcement and spy agencies could see what everyone was doing, they could see terrorists getting ready to commit a terrorist act and stop them before they did. This emotion supported a decade of boom times in the Washington, DC, area as government agencies were established and filled with employees.

Historically, it has been a nice wish, and that is all. Will it still be just a nice wish in 2050? I think so. It is likely to remain a nice wish as long as terrorism gets so much advertising benefit for their cause from successful high-profile terrorist acts. The spectacular example of that happening in 2014 was cyber terrorists convincing Sony Entertainment to cancel showing the movie *The Interview*. At first, it wasn't going to be shown at all, but then Sony partly reversed itself and showed it online and in a handful of gutsy theaters.

As long as terrorists are getting this kind of bang for their buck, they will keep coming up with new ingenious ways of scaring people. Even pervasive surveillance won't stop this.

Surveillance Affecting Governing using Violence

Surveillance won't stop terrorism, but it has already made a big difference in governing styles. The surprising twist concerning surveillance is not the September 11-inspired secret surveillance, but how the more routine and overt forms change governing styles. Surveillance already makes a big difference in how the powers that be and protesters handle violence. Here is an example of what difference pervasive surveillance has already made in how communities are governed.

In the 1920s and 1930s, political movements and governments engaged in a lot of violence to further their causes. Hitler and the Nazis were ready, willing, and able to trash property and people to further the cause. Lenin, Stalin, and the Soviets famously engaged in their full share as well. And these cases were not exceptional. There was deadly violence in the political processes happening all over the world.

Fast forward to the 1990s. In the 1990s, governments and protesters could not use this kind of violence. Marcos in the Philippines, Ceausescu in Romania, Gorbachev in Russia, and Baby Doc in Haiti were all leaders who fell with a fraction of the violence that tumbled governments in the 1920s and 1930s. The violence that ended the Tiananmen Square protests of 1989 became an enduring national embarrassment to the Chinese government, not the business as usual it would have been during the Maoist Cultural Revolution Era just twenty years earlier.

What was the difference? The difference was communications technology.

In the 1920s and 1930s, real-time news, such as a Roosevelt fireside chat, were carefully staged from a radio studio filled with a lot of expensive equipment. Visual news was a weekly news reel shown before movies in a movie theater. These news reels were carefully edited. All this expense and editing meant that those in power had a lot of control over what the audiences heard and saw. This was the era when propaganda was king.

In the 1990s, TV cameras could be shoulder-held and broadcast from on-the-street locations, and citizen-held amateur video was appearing. Rodney King getting beaten up in 1991 was the first famous example of that. Fax and the Internet were also in play, so average citizens could tell the world about events they were witnessing real time.

The difference this made was that violence could now be seen happening in near real time, and neither protesters nor governments could deny when they were throwing fists, rocks, or rubber bullets. And it turned out that the various publics that saw the violence were outraged enough that they wouldn't support the perpetrators. This was a night-and-day change from previous eras.

In 2050, this kind of real-time viewing will be even more pervasive. What difference will this closer surveillance make?

What Will Privacy Consist of in 2050?

Privacy is not well defined in the 2010s, and privacy issues will still rage in 2050. Privacy definitions in 2050 will fall into two broad categories: What is human-human privacy, and what is human-cyber privacy? They are going to be very different.

Take, for example, medical wearables. The cyber these report to are going to have full access. If these cyber programs/entities hear about something that needs correction, they will pass that information on to other cyber and humans who can do the diagnosing and correcting. The issue becomes one of who belongs in that chain of reporting. Who should be able to prescribe, and how strong should they be able to make their prescription? This is an age-old issue that will continue into the 2050s, but be updated. An example of updating: What if people are not as active as the prescriber thinks they should be? What if they have been eating greasy cheeseburgers for lunch every day for a week? What if they didn't get enough sleep last night? In these cases, what action should be taken by the prescriber? Send in social workers or social cybers to intervene? Any at all? Or are these private concerns, even though they can add to the community's health care costs?

For example, a woman is worried about her husband. She wants to know where the driverless car takes him. When should the system reveal this to her? When should it include pictures of him in the car? The system is likely to be smart enough that if there were someone else in the car with him, it could conceal that. When should it do so?

Business is going to be divided into two broad categories: cyber-controlled and artisanal. How privacy is handled in these two categories will be quite different. The artisanal businesses will be conducted much as small business is today, and the proprietors will be concerned about privacy in much the same ways. If a company manager uses pervasive surveillance to discover than an employee is goofing off or gaming the system in some fashion, should he or she be allowed to take action?

Cyber-controlled businesses will be in a completely different world. It will be a world few humans will comprehend, a world centered on lots more coordination and fast feedback. Privacy will not be as big an issue in this world.

What is That Person Thinking?

An eligible man looks across the table at a singles bar. He's been talking with this woman who could become special to him. Should he make his move now? What is she thinking?

A woman is making a business proposal to a potential client. Has she said enough? Is it time to close? What is he thinking?

One of the surprising twists of people being equipped with lots of wearables is providing more and more answers to this eternal question. In business, romance, and domestic relationships, knowing what the other person is thinking is quite valuable. If wearables can help decode this mystery, they become even more valuable.

The early efforts will resemble body language—the questioner will be looking at those things that wearables can monitor such as heart rate and hormone levels—to second guess their thoughts. These early stages will simply be advanced intuition or cold reading. They will be helpful, but there will still be a lot of interpreting needed.

The midlevel efforts happen when the questioner can monitor what the questioned is accessing in cyber space. What research are they doing? What reports are they consulting? Knowing this as well as knowing the emotional reactions to that research will allow much more precise responses if things aren't going the way the questioner wants.

Midlevel is likely to be the top level reachable by 2050. The advanced level is actually monitoring the thoughts that leak into cyber space as a person is doing routine thinking. This won't be easy, in part because thoughts are so fleeting and researching them will take so much time.

How much of this style of researching other people's thoughts will be proper will be a hot topic. A related hot topic will be hacking wearables to do this kind of decoding without the questioned's permission. This will be a red-hot privacy issue of 2050.

Further Reading
- This January 3, 2015 *The Economist* article, "Your phone says: "Cheer up!" Software that senses how you are feeling is being pitched to gadget-makers," is about wearables and phones sensing the owner's emotions.

Top-Forty Jobs

As cyber takes over big business manufacturing and service functions, people will be liberated from the drudgery jobs and free to take on many other kinds of jobs. What they will choose most often will be jobs about which they are passionate.

The surprise here is what kinds of jobs get selected. Jobs of passion are essentially jobs that entertain the person doing them. Because they are entertainment jobs, they will structure themselves in similar ways to other forms of entertainment.

This means we are going to see … wait for it … a top-forty jobs list

The History of the Top Forty

The top-forty phenomenon first came about in the 1950s. What the radio station owners noticed was that while there were hundreds of songs available, listeners wanted to hear just a few—and hear them over and over again. What evolved was top-forty radio.

A 2010s example of this happening is in movies. Thanks to the rapidly declining costs of movie making and displaying, a lot more people can make movies than they could in the past, and a lot more movies can be displayed. In spite of this growing potential for wide variety, the popular offerings remain quite limited in scope. Sequels, prequels, and franchises remain very popular.

I predict that this same top-forty mentality will structure what jobs are popular in 2050. There will be lots of potential variety, but it won't be taken advantage of. Most people will pick from a few kinds of jobs to be passionate about, and they will do them over and over. Very much related will be lifestyle choices. I will talk about them first.

Top-Forty Lifestyles

Lifestyles and job choices are interrelated. People will choose their lifestyle first and pick a job that is compatible. Here are some ideas on what the top-forty lifestyles will be. Much of humanity will live in variants of one of the following lifestyles.

- *The dilettante.* Many people will vigorously pursue one or a handful of activities and get very good at them. This is comparable to what young boomer retirees do in the 2010s, only more so. These dilettantes will be very good at what they do and get praised for it, but their activities won't be contributing much to the prosperity of the community. The activities will be personally fulfilling, and that will be the center of their value. These activities will be high-profile passionate work.
- *The hermit.* These people will live quiet lives with Internet-style social networking interactions being what they spend most of their day at—the guy in parent's basement playing computer games scenario. These people will be numerous, and they will be the backbone support for human entertainment endeavors of all sorts. They will watch and comment, and their support will make or break costly entertainment projects such as the Hollywood movie equivalents of the 2050s.
- *The metrosexual.* A person who gets out regularly but spends most of his or her attention and discretionary resource on fashion, gossip, and clubbing. The artisanal variant will work diligently at getting his or her surrounding environments just right. One variant on this lifestyle will be the successors to the 2010s reality show nouveaux riche. They live to show off wealth.
- *The fashionista.* Fashion, like entertainment, is going to grow steadily in people's awareness. People will spend more time getting fashionable, and they will spend more time paying attention to other people's fashions. The variety of activities that people will engage in to be fashionable will grow as well. An example from

the 2010s is the growth in tattooing and piercing. Another is the growth in cosplay done for fantasy-themed conventions.

- *The crusader.* A person who gets enthusiastically behind a cause. The cause will be emotion-driven. It has to be emotional because in this time of plenty, the harsh reality basics are being taken care of by cyber, so there's no reason for analytical issues to carry oomph. These crusaders will call for the community's discretionary resources to be allocated to solving emotion-driven causes. Guilt will power a lot of crusading. These crusaders will get behind a cause to absolve themselves from guilt they can feel for many emotional reasons. An example would be absolving themselves for the actions of their ancestors. Efforts in the 2010s to fix American racism and injustices to Native Americans fall into this category.
- *The nomad.* These are people who move from place to place. They will be a mix of hermit and crusader in personality. They don't get along well with others, but they will feel they are supporting a cause. The cause will be highly individual for each nomad and highly mutable from day to day. These people will be homeless when they want to be, but there will also be many community-sponsored shelter alternatives available when they choose to use them. Many nomads will be gaming the system in traditional homeless ways. They will be sign holders on street corners and petty criminals, and they will misuse public places. An example of this in the 2010s is using public restrooms as long-duration break rooms. One of the new ways in 2050 may be using public access driverless cars as long-duration break rooms. They will justify this system gaming as the 2050s version of sticking it to The Man. Nomads will symbiote with the social-justice people who want to save them. These will be the people who have a lifestyle that likes to rescue people who look like they are in a tough spot and need help, a variant of busybody/social worker mentioned below.
- *The local entertainer and beggar.* As mentioned in the nomad section above, begging and supporting begging are instinctive ways of thinking and acting. They will continue, pursued by both nomad types and stay-at-home types. A related activity is street entertainment. Related to street entertainment are various forms

of local entertainment. These will all thrive. The entertainment instinct will also support parents supporting their kids in school-oriented entertainment projects.

- *The big-time entertainer.* Large-scale entertainment activities will also be a popular activity. These will be numerous and diverse in both how they are conducted, how they are displayed, and who the audience is.

- *The clannist.* These are people who get deeply into the clannish/cultist lifestyles. There will be lots and lots of variations. These are people who let their us-vs-them and Neolithic village instinctive thinking thrive to form small closed groups. A 2010s mix of clannist and nomad are the Roma who wander in Europe. The root philosophy of the clans will vary widely from communal to religious to charismatic personality worship.

- *The busybody/social worker.* A lot of people will still want to meddle in other people's lives in prescriptive ways. These are the people who are constantly watching for other people who need help to straighten their lives out, who are polluting the moral values of our community with their outrageous actions, and who are endangering the children. I call them social workers because this is the kind of job role people of this inclination will migrate into. Some of these people will support nomads. Other styles will support various flavors of unionism and socialism. The socialist and unionist movements of the 2010s will transform to adapt to 2050 realities and still be a big part of some people's lives. Solidarity is a powerful emotion based on us-vs-them thinking.

- *The system gamer.* Gaming the system brings deep pleasure to many people. Exploiting a loophole or a right to get something for nothing brings a lot of satisfaction. There is a lot of system gaming going on in the 2010s, and it will expand as stable TES get installed, so it will be extensive in 2050 as well. These system-gaming people will have a lifestyle of discovering rights they should have and then demanding compensation for those rights through various kinds of protests. An alternative to this is being an advocate for other people demanding rights. Many lawyers do this in the 2010s. Yet another form of system gaming is corruption.

Corruption in human-human interactions will still be common in 2050. System gaming will mix easily with other lifestyles, so practitioners will also be engaging in one or more of the other styles I have talked about. System gaming will mesh quite nicely with the rise of cyber lying. Cyber can easily act as if a person is winning at system gaming when they are in reality fitting neatly into what cyber have planned. Likewise, discoveries of system gaming are newsworthy as well. Discoveries feed the outrage emotion quite nicely, so discovering system gaming and being outraged about it will not stop either.

- *The quant junkies.* These are people who are deeply concerned with their health and their human performance. They will be dilettantes of human performance. They will monitor and tinker. They will be big on sports and other forms of personal physical attainment. They will be looking for ways to level up in physical performance. They will be constantly running afoul of the busybody instinct in others around them as the question of how much enhancement can be done and still have a pure body (as an example, pure enough to compete in the Olympics). The converse attitude will be run what'cha brung, meaning whatever enhancement you can afford is fine. It is likely that a whole different genre of sports will grow up for those with this second attitude, and they will be scorned and controversial to the purists. Think of the attitudes about pro wrestling and MMA-style contests in the 2010s.

These are some thoughts about lifestyle choices. There will be some variety, but this is top-forty stuff. That means there won't be nearly as much variety as the system could support.

Top-Forty Actual Jobs

Lifestyles are one thing; actual jobs are another. Here are some ideas on what the actual jobs will be in 2050.

Work, Harsh Reality, and Satisfaction

The satisfaction that comes from doing work has a lot of instinct behind it. In hunter-gatherer times, everyone worked; everyone contributed to the well-being of the Neolithic village. This is one condition our instinctive thinking is built to work well in. Another condition, not quite as ancient, is to have royalty of some form who tell the peasants of some form what to do.

But conditions in civilized living are different, and this means that instinctive thinking doesn't match harsh reality as well as it did in more primitive times. Our harsh reality has changed, and our relationship with harsh reality has changed. For example, for the most part, we no longer personally kill, prepare, and cook what we eat. We let numerous specialists and specialized machines transform plants and animals into consumable products. This means our current harsh reality is ordering a Big Mac at a drive-thru speaker and getting it in a paper bag, not raising and slaughtering a cow and foraging to find and root out potatoes to eat with it.

Harsh Reality and Delusion

One of the virtues of human thinking is its adaptability. We can grow up next to a beach on the Arctic Sea or grow up next to a beach on a tropical island and be equally comfortable with our lifestyle. Growing up in a civilized environment is dramatically different from growing up in a primitive Neolithic village environment, but we manage quite well at both.

We manage, but there are dramatic differences in what is acceptable between these many environments. Because we are so good at adapting, we take many of these differences for granted.

One of the differences these changes in harsh reality allow is changes in what is okay for our emotions to tell us. In the above example about food gathering, the civilized environment allows animal rights activists to gain serious community attention rather than be laughed at as strange, hopeless romantics. This September 15, 2013, *Telegraph* article, "Who You Gonna Call? Belief in Ghosts is Rising" by Jasper Copping, is another example. This is about belief in ghosts rising in England.

As mankind's lifestyle has evolved from primitive to civilized, the issue of what is satisfying work has also constantly evolved. We have moved from tilling the land to driving a tractor that tills the land to designing software that makes a tractor that tills the land. Because industrialization dramatically increases the pace of change, this question of what people can do that is satisfying and enfranchising has loomed larger and larger for over a century now … and the looming is not stopping!

Historical Example

The Roaring Twenties was a time of booming economy, booming technology, optimism, and social liberation. It wasn't all pleasant, but it was exciting. And planning for the future was quite different from how it was in the Agricultural Age. Here is an example of the uncertainty all this new technology brought.

Suppose you were in the delivery business in 1920: What skills did you have to learn?

One of the major skills was driving a team of horses to pull a wagon. A second was caring for the horses.

Now, imagine your son says, "Dad, I want to carry on the family business. What skills should I learn?"

If you answer is to learn about horses, your son is going to be in for a painful surprise because in the 1940s, motorized trucks are going to replace horses and wagons. He should have learned how to drive a motorized truck and how to repair it. But how could you—or he—have known? This would have been darn hard to forecast. Jobs of 2050 have that same relationship to jobs of the 2010s.

And there were many other scary exciting things happening besides jobs changing, pleasantly exciting things as well. The book *The Great Gatsby* by F. Scott Fitzgerald is in part a description of that amazement of living in the 1920s.

Then in the 1930s, the whole world experienced the Great Depression, a time when the economic systems that were supporting that 1920s optimism seemed to get mucked up and dysfunctional.

There was a night-and-day different experience for those living in the times. But curiously, during both of these periods, people who thought about social institutions were marveling at the changes the current wave of industrialization were bringing to how people lived. Their thinking and writing changed hardly at all. Asimov's article (in Further Reading below) is a classic example. And now it is my turn to take a 2010s swing at this forecasting.

Building Enfranchisement without Building Stuff

The heart of the issue is what can people do that is enfranchising but not work in the big business manufacturing or service sense. This is the kind of work that automated systems will be handling more and more.

Here are some possibilities I have come up with.

- Entertainment. The core of entertainment is channeling emotion, and this is something humans will continue to do well and enjoy doing. Entertainment and entertainers have been growing steadily in importance to the community as prosperity has grown. This is likely to continue so entertainment-related activities will be keeping a lot of people busy and enfranchised in 2050.

- Fashion. Related to entertainment, fashion will boom as well. As people get more prosperous and more divorced from having to do things for a living, they can devote more time and attention to fashion. In the TES environments, it will boom mightily. Fashion is very much a human-oriented issue. The choices are very human and instinctive, so humans will be at the forefront for a long time. Where fashion is artisanal, humans will be doing the crafting as well. But where the results are mass produced, humans will be handing off inspiration to cyber once the inspiration transforms into something material or serviceable.

Fashion is often about pushing the limits of outrageousness. This is going to continue, and with the new technologies available, the results are going to be astounding by 2010s standards. Consider how astounding the flashing shoes that kids wear would be to the people of 1910. It is not just a technology consideration. Think about how astounding short skirts and shorts would be. 2050 is going to be just as astounding on both fronts.

With this in mind, one fashion form that is going to be interesting to watch evolve is tattooing (and its cousin piercing). In the 2010s, this is very much a human-dominated activity. But as is often pointed out in social media, humans make a lot of mistakes in this realm. It will be interesting to see when and how this process gets handed over to cyber because the results then become faster, cheaper, and more error-free. Plus on the technical side, those flashing kid's shoes will get competition from flashing skin, plus chameleon-like changeability, plus ... mixing narcissism and outrageousness ... wow!

- Creating Human-Crafted Wares: Artisanal Manufacturing. Many people buy something because they feel it has mystical properties. In the 2010s, this is the heart of marketing luxury goods and foods. This market will remain vibrant. Many people will be able to make a living by crafting stuff that will gain mystical properties in the eyes of purchasers because of how it is crafted, as in, it will be human-crafted. This may seem like work, and it may feel like work, but it is not because it really isn't supporting civilization. These human-crafting processes will be hugely inefficient when compared with the automated ways of making stuff that cyber will be managing. As such, this style of making things is icing on the cake. It is luxury. It will be sustainable because the hand-crafting aspect will add a mystical nature to the product. In a prosperous community, many people will be willing to pay extra for that. The buyer will be happy, and the person doing the crafting will feel they are working, which will help them feel enfranchised.

Adding to the handcrafting demand will be a transformation of harsh reality thinking that will also be going on: As processes become more automated, people are less aware of how stuff is really made (as in, the physics, chemistry, and economics of production). This has two effects: The first effect is that people will be thinking the equivalent of, "Why not believe in mystical powers? My harsh reality can support it." The second effect is that because of the human ability to take things for granted, a lot of cyber-controlled automated product can slip into the handiwork. For example, the person making a handcrafted table is not going to be concerned about how handcrafted the glue he or she is using is ... unless worrying about this becomes a fad. The third effect will be that "rights people" will be constantly arguing that artisanal works should be considered part of necessity buying. (As noted in the Allocating Wealth section, necessity buying will be handled very differently from luxury buying.)

- Experimenting. A subset of the human crafting talked about above will be human experimenting. Some humans will keep trying to figure out better ways of doing things at the hands-on level. Much of this will be to improve artisanal crafting talked about above.

A smaller subset will be humans working in with cyber to improve how cyber-controlled products and services are produced. This kind will be comparable to creating the disruptive technologies of the 2010s. This will be work in the traditional, improving-productivity/making-better-products sense. Note that compared to most people in 2050, these people are going to engage in seriously strange thinking styles. They are going to be heavily analytic thinkers, not heavily instinctive thinkers.

- Selling and advertising. People will still buy things, and the buying will be in two general categories: necessities and luxuries. Helping people buy stuff in both categories will be something humans stay involved in, but the products in the two categories will be treated quite differently. The social worker-inclined types will be working with marketing the basic necessities. They will include a lot of prescription in their selling styles. These will be sold to all people, and much of what is sold will be taken for granted by those buying.

Luxury categories will be different. Dilettantes, hermits, and party people in their various incarnations will be large target markets for the luxury categories. These groups will aspire to own things they don't yet have, so there will still be big efforts spent on marketing, advertising, and market research in the luxury items category.

Advertising and marketing will be a human-centric activity in those areas where the results remain unpredictable. As the results get more accurately predictable, cyber will take over more of the activity. For example, the 2010s "One trick to [X]"-style Internet ads will be among the first to go cyber.

Humans will also stay at the forefront where trust is a big issue in the buying choice. Collectibles will be a market dominated by trust. Investing advisors will also be selected on the basis of trust. But the investing process and products are going to be very different. Personal finance will be greatly transformed by 2050, and investing will become a dilettante activity.

- Selling urban legends. Face-to-face selling will remain a powerful way to convince people to buy stuff. One variant of it that will gain in strength is selling stuff based on urban legend. This is because urban legend gets its power from stroking instinctive/ emotional thinking, and that feature of human thinking will be strengthening. Emotional thinking and the urban legends it supports will become progressively more influential as people become more and more divorced from the harsh realities that would prove the urban legends wrong. One example is the antiscience movements that support creationism. These beliefs work just fine as long as you're not trying to solve a complex science problem. Another example is selling wondrous foods and medical cures based on mystical power. These are supported by the deep instinct to worry about food and health. Another example are the animal rights movements. Animal rights can feel quite warm and fuzzy if you're not a person who routinely slaughters many kinds of animals.

- Supporting mythical rituals. I attended the 2013 Salt Lake Comic Con. It was a deeply surprising success. It was the biggest convention *ever* in Utah, and the third biggest Comic Con in the nation. Only San Diego and New York City surpassed it. The attendees were both numerous and deeply into cosplay, designing and wearing elaborate costumes for other people to admire and shoot pictures of.

This Comic Con experience was a vision of the future. This is an updated county fair, and the attendees are getting a lot of emotional reward for their effort. Supporting mythical rituals will occupy more and more human attention as the time and attention spent on work decreases. And as this Salt Lake Comic Con demonstrated, these efforts can bring a lot of emotional satisfaction.

That brings up the question of what mythical rituals are. My definition is a broad one: It is things we do because they make us feel better on the emotional level. To be a mythical ritual, enthusiastic emotion matters, not correlation with harsh reality. This means it includes things such as cosplay and backing sports teams.

- Disaster response. Disasters are always surprises. This means they occur, and responses have to be novel. Dealing with novelty is an area where humans will outperform automation for a long time. Humans will be at the forefront in two areas: First, they will direct the logistical responses to disasters until cyber can catch up. Second, they will provide a lot of emotional comfort to those experiencing the disaster. So preparing for and responding to disasters will remain a highly enfranchised human activity. This is similar to the activity of firefighters and other first responders we experience in the 2010s.

- Military. Humans won't need military, but that doesn't mean it will go away. There is deep emotion supporting a warrior class and being prepared to defend the homeland. What exactly soldiers will do in 2050, I don't know. But they will be around in some form, and being a soldier will be an enfranchising activity.

- Missionary work. Spreading the good word in its various forms is likely to grow as a human-based activity. This activity harmonizes with the chosen people and the helping-the-poor instincts. It is an activity that also supports aggressive hypocrisy and lots of promotion infrastructure. When the good word involves miracles as part of its benefits, many people have to assemble the infrastructure to both document the saints and their miracles and let other people witness them in emotionally impressive ways. Current examples of this happening are the infrastructure that supports the saint industry in India and Nepal and Christian religious revivals in the United States.

These are some examples of top-forty jobs that we will see in 2050.

Further Reading

- This section was initially inspired by an August 30, 2013, *Wall Street Journal* editorial, "Work and the American Character," by Peggy Noonan, in which she discusses how important having a job is to being American.

- I was further inspired by an interesting August 31, 2013, *Open Culture* article, "Isaac Asimov's 1964 Predictions About What the World Will Look 50 Years Later — in 2014," in which Asimov talks about the changing role of work from a 1964 perspective. Both articles point out that as the workplace becomes more automated, the ability for humans to have a meaningful job as makers of stuff diminishes. The machines are doing more and more. I will add to this that in the 2010s, service jobs are facing this same trend. An example of losing service jobs is robots answering phones and making routine calls. A quickly approaching example is driverless cars. In the near future, personally owning a car will become an anachronism. Everyone will use some form of taxi instead. In sum, the challenge we civilized folk are going to face in 2050 is finding enfranchising alternatives to "get a job!"

- This September 24, 2013, *The Economist* article, "Working hours: Get a life," is another about how working hours have changed over the last few decades. In prosperous countries, those hours have declined.

- This January 18, 2014, *The Economist* article, "The onrushing wave," is a nicely done article about this topic.

- The March 29 to April 4, 2014, *The Economist* issue has a special report on the Rise of the Robots, insightful articles quite relevant to this issue.

- If you would like to see a full book based on this idea, check out my book *Child Champs: Babymaking in the World of 2112*.

- This January 27, 2013, *New York Post* story, "One year on the job, 13 years in rubber room earns perv teacher $1M" by Susan Edelman, about an NYC teacher exploiting the system is typical of what I see coming in system gaming. A couple of elements in this incident are noteworthy. First, the teacher is happy to be blatantly gaming the system. Second, the other teachers around him are not-so-happily supporting this gaming because of us-vs-them mentality. In their thinking, the first element is fear of the following scenario: "They came for the pedophiles, and I said nothing. They came for the incompetents, and I said nothing. Then they came for me! To stop this progression, I must sacrifice my common sense to solidarity."

- The second presumption that supports this is that the teachers can't trust their managers. Given the opportunity, the managers will cheap shot them (my term). This distrust enflames their love of prescription, which shows up as loving work rules. (Interesting that in this case, the manager is the government that is there to protect them.)

- This December 13, 2014, *The Economist* article, "The future of luxury: <u>Experience counts</u> Providers of luxury need to offer more than expensive baubles to take advantage of a growing market," is one of a series in this issue on marketing luxury goods. The series talks about how marketing luxury goods is evolving in the 2010s. This kind of evolution will still be going on in 2050. And in those areas where what works remains a surprise, this will still be a human-oriented activity.

- This September 6, 2014, *The Economist* article in its Technology Quarterly Q3 2014 issue, "<u>Biohackers of the world, unite</u>," describes DIY experimenting going on today in biotech. This style of experimenting is likely to continue in 2050.

- Here is an September 18, 2013, *Forbes* article, "<u>Salt Lake Comic Con Sets Record By Leveraging Social Marketing Trends</u>," by Cheryl Conner, describing some details of the convention.

- This January 26, 2015, *Wall Street Journal* article (mentioned earlier in this book), "<u>The Fight to Save Japan's Young Shut-Ins</u>," by Shirley S. Wang, describes the hermit lifestyle already spreading through Japanese culture.

First Responder Blues

Disasters are about the unexpected. This means that humans will stay involved in disaster recovery much longer than they stay involved in the routine and well-planned manufacturing and service activities that are the foundation of big business.

This is a tale of humans staying involved in keeping their community healthy and prosperous by helping recover from the surprises cyber can't respond well to.

They don't call us first responders for no reason. When Mother Nature, humans, or even cyber seriously muck things up and the infrastructure grid stops responding, we are there, Johnny-on-the-spot, to contain the problem and save the people who can be saved. Once the fuckup is contained, the cyber and disaster relief people can get in and restore the grid as quickly as possible.

That's what my mates and I do. We're first. We deal with the unknown and make it known again. When the situation is understood and contained, we get followed by the disaster-relief people. They provide longer-term stuff such as food, shelter, and comfort counseling. And the cyber-repair crews put it all back together again, good as new.

That's what happens. Sometimes, it is simple and straightforward— like some whacked-out arsonist torching an abandoned house—and then there are the other times.

There was the time we responded to a kitchen fire. We got there and came in the front door, and sure enough, there were flames in the kitchen. But when we put water on them, they just lapped it up like a thirsty dog. The chief saw this and said, "Whoa! Everyone out!" We hustled out and, sure enough, four minutes later, the entire house collapsed into a swirl of flames. It turned out what was burning was the entire basement; it was just coming out through the kitchen. I'm sure happy the chief was as experienced as he was and on his toes when we went in. The house, by the way, was off-grid by then, of course, so the cyber didn't have a clue, either.

These are the kinds of surprises we have to deal with.

That house one was small. On the big side, there was that volcano erupting a couple years back. That was a fine mess. The ash cloud was collapsing roofs and dish towers left and right, and the ash in the air was screwing up radio communication. That was the low-down ash. The high-up ash was mucking up airplanes and drones as well as visibility. No one could see what was happening.

We first responders were in there with our suits and gas masks getting people out. The big worry while we were in there was whether one of those ash cloud eruptions would turn pyroclastic, come roaring down the slope, and burn and blow us all away. The observers were watching for that. It was no walk in the park, I'll tell you. But we saved about two hundred lives that day. I think it was worth it.

This latest one is a doozy, but in a totally different way. It's not Mother Nature in our way; it's people. There's this whacked-out, back to nature, cult group living in an abandoned theme park west of here. The trouble started when our boys came out to do a routine inspection of its building monitoring system.

The cult leader responded with, "Building monitoring system? We don't have one of those. It is part of the evil of modern civilization."

The inspector said, "That may be, but if you have a building in this county, you have to have one."

"It wasn't here when we came."

"That's because this place was abandoned 20 years ago. Times, and regulations, have changed."

"Well … we aren't living in the buildings. We are living in tents outside them."

The inspector looked over the cult leader's shoulder. It was a blatant lie, but it sounded like a good, "Yes, but …" to the cult leader.

The inspector made a note in his communicator and then asked the cult leader, "Do you have a communicator?"

"No. They are evil."

"Okay. I've issued you a warning. I'll have the office send you a hard copy. You have a week to get this fixed."

And the inspecting team headed back to the office. This having humans do inspecting was a pretty rare event. The drones had reported new activity in the area, which is why they came out at all.

It turned out that a building-monitoring system was far from the only essential not in those buildings any more. Another was water. Those weirdos were hauling in water from a well. That was crazy, but building the camp fire inside one of the buildings was even crazier, and, sure enough, it got out of control. We were back two days after the inspection putting out a blaze that had spread to three buildings. We had to pull twenty cult members out of that inferno. It was pure luck for those cultists that a nearby crop drone had reported a screwy-looking forest fire and that report had gotten to us quickly.

Whew! It is just crazy what some people do these days. Here's what I have to say about that particular situation: If you want to be a back-to-nature cult member, grow up on a dirt-poor farm. That way, you'll have some common sense about dealing with "mom".

Child Raising

Child raising is a highly emotional activity. Many people will devote a lot of effort to having children and raising them in the best way possible. Child raising will consume a lot of emotional attention in 2050. As a result, it will suffer a lot of the curse of being important.

But the total number of human children being raised by humans will be small. It will be too small, in fact, to sustain the human population. The surprise of 2050 is the big challenge of getting enough children raised. The curse of being important will swirl deeply and widely as this challenge is faced. This means that there is going to be some diversity in how children get raised and lots of heated emotion swirling around in on-looking people wondering if the people they see doing the child raising are doing it the right way.

One of the blind spots in 2010s thinking is not recognizing that prosperous urban environments are a population sink, not a population source. An example of shrinking population due to prosperity is Japan in the 2010s; another is Europe. The cause of this is that people living in prosperous urban environments have lots of distractions that reduce their time and interest in child raising. The classic example of this is the career-or-family question, but that is just one of dozens of distractions.

Because there is so much technology available and so much emotional interest in the activity, child raising in 2050 is going to be a diverse activity, and done for diverse reasons. That is what this essay is about.

Many Choices Available

This business of child raising in the future is at the core of my previous Technofiction book *Child Champs*. It is something I have given a lot of thought to.

One of the realities is that there will be lots and lots of choices available for creating and raising children. But the hard question is going to be magnitude—how to do enough of it? There will be a strong tendency for humans to take a gourmet attitude toward child raising. They will want to do a fine job on just one, maybe two, kids. This will then leave as a chronic question of how to fill the baby gap. This is the difference between the number of kids humans want to raise and what is needed to sustain the population.

The Prosperous Urban Distraction

This baby gap is not a new problem. Cities have been population sinks ever since there have been cities. But when the human population in cities is a small fraction of the total, it doesn't matter. But modern prosperity is changing the equation. Around 2010, half the world's population became urban; the percentage is continuing to rise steadily. This means that this baby gap is already transforming from a curiosity into an issue. If the population becomes ninety percent urban, which is likely in 2050, it becomes a high-priority issue. This is not speculation. The current United Nations population projections call for a peak around the 2050.

As stated earlier, this fertility change is not due to advances in birth control technology. It dates back centuries. It is due to distraction. There are so many things for urban dwellers to do besides raising babies that it gets pushed far down the priority list, and fewer babies are born and raised. For example, in Europe of the 2010s, the fertility rate is 1.59 children per woman. A rate of 2.1 is needed to maintain a stable population. Places such as Europe maintain their population through immigration from rural areas in other parts of the world. When those rural areas depopulate— because all the young people have moved to the cities—the crisis hits with full force. It has hit in Japan already because Japan has long discouraged immigration.

Who Will Make Babies in 2050?

Who will make babies in 2050? Here are some baby sources I envision being active in the 2050 environment.

- *Gourmet babies.* Many women will produce children in conventional arrangements such as marriage. Many couples in alternative coupling arrangements will also do conventional baby making. All these styles will also engage in alternative child raising such as traditional adoption and something up and coming: buying designer children.

The alternative couplings will be much more common in 2050 to avoid the legal and fiscal minefield that divorce has become and likely still will be. This toxicity has grown over the years because of instinctive thinking. The scorned person (I sacrificed for you, now you owe me ... big time) mentality is instinctive thinking that is shared by both the people involved and the community around them. And like much instinctive thinking, if harsh reality is not constraining it, it gets crazier as prosperity grows.

What these various coupling styles have in common—what makes them gourmet baby raising—is being ready, willing and able to commit a lot of resource and attention to the child-raising process. These kids will have the best their parents can afford and lots of parental attention. This is why I call them gourmet babies. There will be a lot of these, but not enough to sustain the population.

- *Baby-crazy people.* A growing source of kids will be people who have a powerful instinct to have lots of children, and they let that instinct flower, even without a significant other to help out. In the 2010s environment, these are the single moms. In the 2050 environment, single moms will get a lot more social attention and support. They will be part of the mainstream. Single mom clubs for raising children will become common. and they will have a lot of social influence. A few moms will go it alone, but going it alone will mean lots of cyber support rather than lots of human support, not no support at all.

- *Be-fertile cults.* How humans gather into groups will be diverse. Some of those groups will be devoted to spawning and raising children. A 2010s example is the Fundamentalist Church of Latter

Day Saints (FLDS) groups. These are fundamentalist religious groups that are both family-oriented and oriented to having lots of children in those families. Many of these be-fertile cults will also support deeply alternative ideas. The FLDS are also polygamist. In 2050, these groups will not be numerous, but they will be around. As with single mom clubs, these groups will get lots of government support for their child-raising activities. The need to child raise will encourage governments to overlook the other activities that these groups engage in that social prescriptionists don't like.

- *My little tax deduction.* In 2050, many communities are going to recognize this baby gap problem and ask their governments to do something about it. One of the quick-and-easy ways is to offer tax breaks. With time, and the crisis growing, and income not meaning what it does in the 2010s, many other kinds of incentives will also be offered. Those who take up these various kinds of incentives fall in this my-little-tax-deduction category. They are having and raising kids because the government is paying them in various ways to do so.

Many of these are going to be single moms, as mentioned above, and one of the new social structures we will see emerging in the 2050 environment is single moms clubs of various sorts. Some of these styles will be spontaneous, and others will be government-sponsored groups who get more benefits in return for following government-recommended ways of child raising. Done well, these single mom clubs can produce good results. This style can be built upon instincts which support a sea of child raisers raising a sea of children.

- *Genetic engineering to produce better babies.* One of the chronic white-hot topics will be how much genetic engineering to apply to fetuses. The instinctive thinking on this will be much like the instinctive thinking on how much drug-taking an athlete can do and still be pure enough to compete. The curse of being important will be strong here, which means lots of opinions, lots of regulations, lots of regulation dodging, lots of misinformation, and lots of scandals. This will remain a very human topic, one that gets a lot of time in the afternoon dramas of 2050.

- *Genetic engineering to produce specialty people.* As humanity and the cyber community master the skills of genetic engineering, new species of humans can be created that will thrive in conditions too extreme for regular humans. An example is low-gravity conditions. Long term, the low-gravity environment is hard on human health. It causes problems such as calcium loss in bones. If genetic engineering can overcome these problems, space-faring people who are not bothered by years of low gravity can thrive. These engineered kinds of beings will be humanlike, but not fully human.

These people will be different. This brings up questions like will humans raise them or will cyber creations/muses raise them? Will they be conceived in normal human mothers? Will other conception and fetus-raising tools be used?

As these engineered styles of people get built for more extreme environments, such as living on Mars or the moons of gas giants, humans can do more and be more innovative in these environments. These engineered people can contribute more to the spread of humanity through the solar system.

- *Vat babies and cloning to produce regular people.* Those people on earth who get into extreme gourmet will want to use all the tools available. This is the tiger mom feeling on steroids. These people will use more fetus-growing environments that are not a human uterus and child raising that employs a lot more creation assistance.
- *We-just-need-people people.* Even with all of the above, the world will need more humans if population is to be sustained. When a community, human or cyber, thinks more humans are needed, these necessity humans will be the humans who fill the gap. These babies will fill the gap that immigrants do in the 2010s. They are being produced to service a need that regular humans or cyber can't. Given that TES provided by cyber is filling basic human needs, what these people will be needed for is dealing with surprises of some sort. At this stage, I can't forecast what the surprises will be.

This is where racial discrimination in 2050 will center on. And many of the traditional ways of thinking about race will reappear. For example, should these people be allowed to marry with "real people"? One consistent cultural difference between these kids and the adults they grow up to be is that they are not going to be pampered like any of the other groups are. They are warm bodies grown to service some need that regular humans aren't filling. Fitting them in socially is going to be a challenge.

The Sex Ratio

One of the most visible choices in child raising is whether to have a boy or girl. As birth control and abortion became widespread, this transformed from a wish into a choice in more and more cultures. Since the 1990s, this has shown up in East Asian cultures as skewing toward more male children, and in the 2010s, it is affecting marriage choices there.

This is going to be an even easier choice in 2050, and just one of many that a child raiser makes in picking the best fetus for them. The big question for 2050, a much more feminist world, will be what ratio gets picked? For example, are single moms and single-mom clubs going to want to raise mostly girls? Is this ratio going to completely flip-flop from what South and East Asia experience in the 2010s?

If so, this is another surprise cultural change that will hit many of the TES cultures.

Tiger-Mom Kids

Another group that is going to have a hard time fitting in are the tiger-mom kids. The parents who are trying hardest and most successfully to have their children make meaningful contributions to society will keep in mind the prime human virtue of adaptability. Cyber will excel at handling routine and switching between various routine activities, while properly trained humans will excel at handling surprises and things that are beyond routine.

For humans to be good at this adaptability virtue, they must be properly trained. They must exercise in it long and hard. This means that the good child-raising practices for enhancing innovation and adaptability are not going to be the ones that support the tender snowflake child raising that works well for raising TES-comfortable adults. This means there will be a culture clash. It will be similar to the homeschooling culture clash of the 2010s. The curse of being important is going to make this a high-profile culture clash in 2050.

Fitting into the Social Environment

All of these children, spawned in different ways and raised in different ways, are going to fit into the 2050 lifestyles. The year 2050 is going to support a lot of diversity, but the foundation will be the various TES lifestyles, and most of those will be urban lifestyles.

Gourmet children will aspire to meaningful lives that involve being more tangible about accomplishments than average the TES thrivers—they are being taught to be movers and shakers as adults. These tangible achievers will be a mix of dilettante and disruptively innovative. The dilettantes will concern themselves with artisanal activities and be liberal arts in their thinking. The disruptively innovative will be fiercely analytic thinkers who do a lot of coordinating with the cyber community so they can help improve cyber-dominated manufactures and services. Those who are raised for other environments, such as space travel, will be analytic thinkers as well.

Those who are raised as tax deductions and fertility cult members are more likely to be instinctive thinkers. Their reality is not going to be manipulating harsh reality in serious ways, so they can let their hearts be their guides. Entertainment and sports, as both performers and spectators, will be the center of their lives. Supporting causes du jour will also be a common activity.

And for almost all these people, wild mind altering will not cause much harm or disadvantage. In fact, it may bring advantage if compatriots start admiring them.

Conclusion

These are some thoughts about child raising in the 2050 environment. The instinct to give children the best is strong, so all kids will get the best their parents can come up with. One of the big differences will be just that—what the parents can come up with.

Raising styles will also be diverse. There will be parents in the traditional sense, parents in alternative senses, single-mom baby clubs, and fertility cults.

The adults who grow out of these styles will be different in their outlooks and thinking styles. As a result, the diversity in lifestyles and thinking styles will remain large, but the range will be quite different from that experienced by the humans of the 2010s.

Further Reading

- For more thoughts on human adaptability, try this December 11, 2014, *Wall Street Journal* article "Artificial Intelligence Isn't a Threat—Yet," by Gary Marcus.
- This April 19, 2015, *Daily Mail* article (also mentioned elsewhere), "Why men won't get married anymore," by Peter Lloyd, describes how marriage rates are declining because the married environment and the divorced environment are becoming so toxic for men. This trend is likely to continue.
- This April 18, 2015, *The Economist* article, "The marriage squeeze in India and China: Bare branches, redundant males," talks about how the high ratio of male children over the past decade is affecting marriage.

The Role of Violence

Violence has been with life on earth since the beginning of animal species, well before humankind appeared. Like many other activities humans engage in, its forms change when the tools and wealth available to the community change. You don't get armies until you get agriculture, you don't get killer drones until you get integrated circuits. Likewise, which forms are acceptable and which forms are ... well, outlawish ... change constantly and from one community to the next.

Pervasive Surveillance

The coming of pervasive surveillance is going to once again change a community's relations with violence. Cameras can show violence occurring, but that is just the start. Wearables can also experience violence. If people are subjected to sudden acceleration, was that from a blow? If their heart rate and hormone levels change to fight-or-flight status, are they being threatened? Are they getting ready to threaten someone? There are lots of ways pervasive surveillance can indicate when violence can happen soon or is happening.

In the same vein, if a restraining order is issued to keep two people apart, with wearables and GPS, this can be easily supported and in a flexible fashion.

Pervasive surveillance is going to dramatically change the acceptable forms of threats and violence. How it will change them will be surprising. Some people, in some situations, will still want to express themselves in ways that are considered violent to some other community members. What is considered violent will not be agreed upon by all community members. Think of the newly evolved trigger warnings in the 2010s. With that in mind, how the lines are drawn and how people will skirt those lines will be quite different in 2050.

Social Media Violence

This is the old ugly-rumors-and-chatter sort of social violence updated by modern communications techniques. In the 2010s, the emergence of social media trolls became a new way to upset other people, and this evolved to have a gang-up-on-someone variant. This problem was amplified by tender snowflake child raising that brought with it heightened sensitivity to trigger-language issues.

These increased-sensitivity-to-criticizing-language trends are supported by instinctive thinking, so unless harsh reality or pervasive tolerance training intervene, they will be even stronger in 2050.

Domestic Violence

One of the surprise changes that cyber muses will bring is that cyber muses can substitute for humans as targets of domestic violence. If humans come home drunk and frustrated, they can slap around their bitch cyber muse to get some relief. (Icing on the cake: These cyber muses can be programmed to piss and moan when the person comes home late and drunk, "so they deserve it." Keep further in mind that the muse can be designed to take this abuse with no harm done. It can be as routine for the muse as talking is.)

How acceptable this cyber-is-the-target style of domestic violence is will be will be up to the community, so it can vary from as acceptable as video game violence (just fine) to as inappropriate as hitting a human. If a community loads up with prescriptionist, social justice-oriented humans who feel in their hearts that cyber muses should be treated like humans, it won't happen as much, but some other practices that allow humans to let off steam will replace it. If the community becomes an across-the-tracks community that tolerates this (in the eyes of the prescriptionists mentioned above), it will happen routinely.

As mentioned above, when wearables can detect a person feeling threatened and report it to other entities in real time, this will affect human relations of all sorts. This becomes the ultimate form of trigger warnings.

The not-so-surprising twist from this is human-human relations in 2050 will be delicate indeed, and in response, humans will prefer dealing with cyber muses even more.

Another response may be to organize *Fight Club*-style gatherings where socially forbidden activities are openly engaged in. ("The first rule of Fight Club is: you don't talk about Fight Club.") How these will mesh with pervasive surveillance will have to be worked out.

Gang Violence

Young men forming gangs is instinctive behavior. It happens when there are no bigger community acceptable goals and organizations for the young men of a community to join and pursue, and they get frustrated. In the TES environments—with their lack of linking survival to disciplined behavior—this lack of compelling bigger goals will be common. Finding attractive, acceptable goals (the next big thing [NBT]) (my term) will be a big challenge for all TES communities. Those which can't come up with popular NBTs will have thriving gang cultures.

But pervasive surveillance will also be in place. So what gangs can get away with in the way of antisocial, law-breaking behavior is something gang members will be constantly experimenting with, and they will find ways. This will be a chronic arms race in all communities with gang problems.

Rebel-with-a-Cause Violence

Closely related to gang violence will be those who commit violence in the name of a cause. This can range from protesters who get modestly violent at a mass protest to the terrorist equivalents of 2050 who commit heinous violence in the name of a cause.

Protesting-related violence is promoted by frustration, just as gang violence is, so it will be just as big a challenge to control in the TES communities of 2050. It is likely to be the era's biggest and most organized violence challenge.

In the 2010s, terrorist violence is encouraged by media pursuing if-it-bleeds-it-leads reporting policies. Terrorist violence is about promoting a cause, and the heavy media coverage does a fine job of promoting it. If reporting techniques of 2050 are clever enough to have ways of reporting terrorist violence without promoting the cause, the amount of terrorist violence will be low.

War

How war is conducted constantly changes as war-making technology and political circumstances change. The famous military truism concerning this is that an army is always well-prepared to fight the previous war, not the next one. This means that the wars of 2050 are going to be nothing like the wars of the 2010s.

This means that the question of what human soldiers will be doing and the kind of violence they will be engaging in is going to have surprising answers. What will human soldiers do when drones, creations, and avatars are doing most of the actual violence and most of these are being controlled by cyber?

In 2050, soldiers on the advanced technology side of a conflict are going to be more humanitarian types than shooting types. They will follow up after the cyber and avatars have done the violence and offer succor to the survivors. They will be offering the human touch to reduce the anguish of the survivors.

The impoverished technology side of the fight may still have humans doing violence to other humans. But given the revulsion that most people have toward violence as well as the pervasive surveillance, these impoverished human soldiers are going to have to be real sneaky about the violence they conduct or they will lose support for their cause.

All of the above said, the powerful instinct to support warriors will still be in place. Even if the soldiers spend almost all of their time marching back and forth, they will still be a well-recognized and well-honored part of most communities. And they should still have lots of "toys" to play with. Defense budgets will remain large.

Mind Altering

Mind-altering practices in 2050 are going to be wild and wooly compared to the 2010s. There will be many more kinds; wearables are going to increase both the variety and the surveillance; the TES lifestyle is going to divorce mind altering from bad consequences even more than the lifestyles of the 2010s do.

And there will still be strong community opinions on what are right ways and wrong ways to do it. So mind altering will still be a white-hot topic in 2050.

Wild and wooly, indeed!

Dangerous Activity? We Don't Need no Stinking Dangerous Activities!

One of the deeply compelling reasons to limit mind altering has been that humans operate equipment that can maim and kill, one of the highest profile examples being driving a car. In 2050, cars will be driverless, and most manufacturing and service jobs will be humanless. In such circumstances, where is the danger? What dangerous equipment will humans be operating? (Work won't be dangerous; but many people will still be engaging in dangerous hobbies. There will still be some bleeding for media to report about.)

Those fully acclimated to living the TES lifestyle are going to be spending most of their time at gaming, partying, entertaining (on both sides—being audience and producers/performers), and dilettante activities. Of these, only a few styles of performing, sports, and racing will be activities that have to be inherently dangerous—the difference between playing a piano and walking a tight rope. In such an environment, mind altering can become much more extensive and intensive than what is experienced in the 2010s. Again, no need to worry about DUIs with ubiquitous driverless cars.

Wearables Will Add to Mind Altering

Many wearables that will deal with health can easily be adapted to mind altering. This will be the new frontier of mind altering. There will be new designer chemicals, yes, but they will pale in comparison to the speed and variety of effects that wearables can produce. Plus, it is likely that wearables will be able to turn effects on and off more quickly than ingested chemicals. They will also be able to feedback more sensitively, so the effects are more what the user wants.

Everyone is High

The result of all this innovation is that most people will be spending most of their time in a mind-altered state.

The social/community questions of 2050 are which are acceptable states and which cross the line. This will be a chronic hot topic because the curse of being important rages strongly in the mind-altering arena. Everyone has an opinion on what is right and what is wrong, and many people get both evangelist and prescriptionist about their opinions. The result is that there will still be illegal ways to mind alter that are popular with a lot of people.

These opinions of acceptable and unacceptable will differ from community to community and may become one of the big differentiators between communities. In response, people will consider a community's mind-altering choices when they are choosing where to live. Just as in the 2010s where people consider the quality of schools when they buy a home, in 2050, people will worry just as much about the mind-altering standards of a community when they look for homes or other lodging styles. And this will come up when they are looking at job choices: What will be a compatible mind-altering work environment?

Mind Altering and Disenfranchisement

There will still be strong incentives for mind-altering abuse in 2050. Even though cures will be numerous and effective, a lot of people will participate in abusive mind altering. The abuse will fall into two general categories: occasional and chronic.

Occasional abusive mind altering will happen because some people want to experiment, push the limits, and see what happens. Mind altering is a simple and convenient way of doing this. This kind of mind altering will happen a lot because there will be so many choices and so many new choices, but the choices will not be repeated by the same person too much. The person will learn and move on. Given how efficient cures will be, hard-to-cure addiction is not going to be a common problem. This kind of abuse will be faddish.

Chronic abusive mind altering will have different roots. One of the important ones is feeling disenfranchised. The participator in chronic abuse is going to be feeling that the community cares little about their opinions and what he or she does is not important to their community. Abusive mind altering is an easy way to act out the frustration that comes with feeling disenfranchised. Solving this kind of abuse means solving the disenfranchisement issue the person is feeling.

News Reporting

News is about timely topics that matter to people. In 2050, what matters will have changed a lot, so the topics that will be reported on will be quite different. Changes in communication technology will also bring about big differences in what are considered timely topics, how they are reported, and what are considered proper actions to be taken by news makers.

A Quick History of News

Being interested in what's going on around a person is instinctive. It has always paid big dividends. But what is interesting changes with the environment. In the Neolithic-village environment, what is interesting is what is happening in the plants, animals, and people around the village a person is living in. In the Agricultural Age environment, it is what is happening in the fields and people around the farm. In the Industrial Age environment, it is what is happening in factories and markets served by the companies a person works for. And in all of the above environments, threats of violence are top priority. In 2050, what will be of interest will have changed again and how these news topics will be disseminated will change as well.

Topics of Interest

Since the people of 2050 are going to be mostly TES-acclimated (as in, divorced from the harsh reality of both working for a living and understanding how things work), people-oriented, emotion-based topics will dominate most forms of news dissemination. This means that news will be about things such as entertainment, sports, human interest, and cute pets. Fearful news will also still be a regular news topic. Such items will still be very much in the news. But what will be causing those fears will have changed.

Mixed in with these will be artisanal-level how-to advice on topics such as cooking, travel, healthy lifestyles, and getting ahead. This kind of information will service the dilettante community members. Another large category will be the spread of gossip and urban legends.

Since cyber will dominate large-scale manufacturing and service activities, business news in its 2010s form will be a small specialty niche. The human who can understand large-scale business activity will be the rare person, so it won't pay to produce much of this sort of news. For the same reason, there will be little interest in hard science and lots of interest in urban legend-based pseudoscience.

How It Will Be Presented

How news will be presented will change dramatically once again. In the 1800s, the high-tech news distribution method was newspapers. In the 1930s, high tech meant radio and movie theater news reels. In the 1950s, TV moved to the forefront. There have been constant revolutions in how news is disseminated. In 2050, things will be different again. Most news will come through wearables. This brings the potential for even more vividness and timeliness than what is possible in the 2010s. How news is interpreted (how a newscaster or commentator will add to this news stream) will be done in new ways as well.

In most cases, this expert opinion will be adding less insight than older systems added simply because things are happening so quickly. There will be so little time for commentators to reflect on what has happened because they will be asked to comment while it is happening. They will be much more like sports announcers are in the 2010s. In addition, because these commentators are as TES-acclimated as their viewers, the opinions will be much more emotion-based than in the 2010s. This means that news will be even more for entertainment than a tool for assisting decision-making.

The Difference the Tech Makes

These revolutions do more than just move news faster. They will also affect how people feel about the actions taking place. This means that what is acceptable behavior will change.

For example, one of the big differences that the 1990s communications revolutions made was making large-scale political violence unacceptable. Compare the violence of the political movements of the 1920s and 1930s with the violence of the political movements of the 1990s and 2000s. The fall of the Soviet Union in 1991 was much less violent than its rise in 1920, and the Chinese government still considers the violence of Tiananmen Square in 1989 an embarrassment, -- which is so different from the feelings about the violence that happened in the Mao-era Cultural Revolution from 1966 to 1976.

What actions—both domestic and political—people of 2050 consider acceptable will be quite different from what is considered acceptable in the 2010s. There will be surprises, of course, but some patterns of the 2010s are likely to strengthen. One is prescriptionist-controlled speech.

"Say It Right, Son!"

Speech intolerance is likely to strengthen because so little conversation will affect the harsh reality around a person. If you're not building or repairing a car, then straight talk about how to do that well is not going to make the world around you work better. This means that if someone comes up with a prescriptionist opinion about car repair that is quite crazy, it makes no difference … unless they can ostracize someone else for not having the same opinion.

This means that emotion can dominate even more than in the 2010s, so prescriptionist opinions about what is proper talk will be much more common in 2050. Correct speech will be on every newscasters' mind in 2050.

Wilderness 2050

Prologue

This is a story about how Americans will visit their pristine wildernesses. To keep them pristine, only one person a year will go in and experience the wilderness in person. Meanwhile, millions of others will experience the visit vicariously through that person's personal sensors.

Chapter 1

The great pine log doors are stained the traditional deep brown of the Park Service. They are brass-bound, lichen-covered, and flanked by towering cottonwoods. Inscribed deeply into the milled wooden beams is the following.

US Park Service
High Uintas Primitive Area

Underneath on a plastic sign hammered to the beams are the words:
--DANGER--
US Park Service Wilderness Area
Unauthorized persons within park boundaries are illegal
AND SUBJECT TO LIFE-THREATENING HAZARDS
No trespassing. Convicted trespassers are subject to
a maximum fine of $1,000,000 and up to five years in prison.

Out of place is the hustle and bustle in front of the gate. There are two cinemobiles, six SUVs sprouting satellite link-ups, an antique Humvee, and ten late-model ATVs. Overhead, keeping carefully outside the park boundary, are two large drones and the Goodyear blimp. Inside the boundary, drifting lazily on the breezes, are two hawks and two Park Service drones. The media folk have clustered an almost-respectful distance from the gate and are being held at bay by a dozen rangers.

The oddest vehicle is the one in the center of the cluster: a black limousine pulling a horse trailer. Out of the limousine and into the center of all this commotion strides Big Jim Steed. Jim is six feet tall and ramrod straight. His weight is hard to tell under all his clothing and supplies, but he moves dexterously without stumbling or a hint of lugging. As he approaches the gate, he doffs his helmet for the cameras. His hair is dark and neat, his nose straight, smile wide, and teeth white and even.

As he checks his equipment, an announcer drones in front of him.

"This is Chet Bradley. Welcome to Wilderness 2050. This year, we will be visiting the High Uintas Primitive Area. An unusual choice, one that hasn't been visited since 2039.

"This ecopreserve, like all the others in the Primitive Department, is set aside to be untouched and undisturbed by people." Chet frowns slightly. "I mean … beings of homo sapien descent. It is large enough to be a self-sustaining ecosystem, over five hundred square miles in this case, spanning from Interstate 80 in Wyoming on the north side through the foothills outside Duchesne, Roosevelt, and Vernal in Utah on the south, and from Flaming Gorge on the east to Kamas on the west. The mandate of its caretakers is to keep it as pristine, untouched wilderness, and they take this mandate seriously. To fill you in about this policy, we've invited Richard Moonan, director of the Park Service, to say a few words."

Moonan, resplendent in his Park Service best, continues. "Thank you, Chet. To keep our American wilderness pristine, only one person may enter any of the 364 Primitive Departments each year. To be fair about it, that person is selected by national lottery. The person is selected a year in advance so he or she may undergo the intensive training it requires to enter a wilderness."

"Thank you, Mr. Moonan. This year, that being of homo sapien descent is Jim Steed, who is standing behind me. I will now see if we can take a moment of his time while he's busy with his last-minute preparations to give us his feelings about being the one person … being … who will enter this year.

"Jim, this certainly must be an exciting moment for you."

"Yes, Chet, it certainly is. I'm honored that I can represent the American people and excited that I'll be the one to actually see the wilderness in person this year."

"Jim, how did you make your selection of the High Uintas?"

"Well, Chet, there were a couple reasons. The High Uintas offer some exciting terrain. It is the only place in the United States that has a whole plateau of terrain at about 13,500 feet, not just a few peaks. Second, it hasn't been visited since 2039."

"That's eleven years."

"That's right. And third, I was born near here and raised in Vernal, a town that's situated on the south boundary." There is scattered applause and cheers from the circle of on onlookers. "I used to sit on the backside of Red Mountain and stare up at the Uintas, dreaming of this day."

"Well, congratulations, Jim. There we have it, folks. A dream come true. Let's let Jim get on his way."

Jim does get on his way after four more interviews with other headliners of the media pool. The Olympics are big, but the Wilderness Event is all-American; the anchors love the patriotic overtones.

In between interviews, Jim consults with Olson, a gray-haired, medium-height man in his fifties with a slightly bent body, simple clothes, and a leathery face.

There is another flurry as Celeste, Jim's horse for this adventure, is taken from the trailer. "Celeste. How did you come up with that name for her, Jim?" asks one of the anchors.

"I meditated, and it came to me," says Jim as he throws the high-tech saddle on her back and adds the other high-tech rigging. Last to go on is her wilderness skirt, a device designed to simultaneously wipe out her tracks and catch her droppings. The horse's natural motion will spread the droppings over the skirt and air-dry them so they can be used for the evening's fuel.

Jim dons his helmet and leads Celeste to the heavy door. The honorary gate keeper approaches. Applause rings out for the still-glamorous Jennifer Lawrence, assisted by Bradley Cooper. Jennifer officially checks Jim's credentials while the press watches on.

"Good luck, Jim." Jennifer hands over the three beepers.

"I'll be back," replies Jim.

The beepers are transponders that authorize Jim to be inside the wilderness boundaries. There are three just to be sure that one will always be working. Jennifer checks to make sure Jim's own wilderness skirt will adequately brush out tracks in soft earth, and she opens the massive door and Jim strides through beside Celeste. The door slams shut with a deep boom, and Jim is alone in the wilderness.

Chapter 2

Jim is knee-deep in grass. Celeste looks at it somewhat tentatively. Ahead of him, to the north, tower the Uintas. As viewers across the land share the view coming from Jim's helmet, Chet Bradley speaks softly. "He could have chosen the historic way in from Kamas on the west side, or he could have followed remains of Indian trails that come in from the east side. He might have canoed part way in through Flaming Gorge and then picked an abandoned highway that penetrates to the high plateau, but he chose the steep approach, up from the town of Duchesne in the south. This will take him through the widest variety of ecological zones on his trip. He will proceed north up Yellowstone Creek to King's Peak and then come out, heading west down the Provo River and Beaver Creek to Kamas."

Jim fastens one of the beepers to Celeste. He is about to mount her when he hears a faint hum coming from behind the gate. Over the top comes a floater drone with lenses and microphones bulging out every which way.

"Tinkerbell," Jim mutters. "They got her fixed in time." Jim mounts up and rides. Tinkerbell, another video- and audio-feed for the rest of the world to experience Wilderness 2050 with, flits around behind him.

Twenty kilometers from the gate, Jim camps in a large meadow. The ground is unnaturally flat. Satellite pictures show the area crisscrossed with rectangular variations in vegetation.

"You can't see it from the ground, but there used to be a town here," Chet says. "The wilderness restoration project has been very effective at wiping out all small-scale signs of human habitation. Only large-scale, low-contrast features such as roadbeds remain detectable and only when viewed from a distance. Jim will be unaware of this former town's existence unless he's been checking his history maps."

At camp, Celeste finally shows interest in eating this long, green stuff growing up from the ground.

"Control," asks Jim, "where was this horse raised?"

After a pause, his Wilderness 2050 contact has the answer. "She's from Kentucky, one of the finest racing lineages there."

"She acts like this is her first time seeing grass," Jim comments.

"It may be. She was raised specially for this trip in a climate-controlled barn back in Kentucky, but I'm not sure how she was fed."

Jim continues scraping dried droppings from Celeste's wilderness skirt. As he scrapes, he thinks, "Christ, this horse has as little experience at this as I do! I spent my wilderness simulator hours on ATVs and flyers. Horse mode was too slow and unpredictable to be any fun. It wasn't until I won the lottery that I even thought about anything slower. I was going to walk this trip, but six months ago, my agent pulls a coup. He argues that this is an adventure in the mountains of the Old West where horsemanship has a long tradition. It demands a horse. The media backs him up. The last couple of years are remembered, trying to wring adventure from watching a middle-aged earth-muffin admire wild flowers in Cape Cod National Seashore and an amateur entomologist stalk a dragonfly nymph. Then he wangled a deal with the horse breeders. Next thing I know, I've got a horse on my list of acceptable equipment. I remember the agonized look on Olson's face as he tried to figure out how to fit horse riding into the schedule."

The next day, Jim enters the Yellowstone Creek gorge. The day is relatively uneventful. He rides and leads Celeste along an abandoned road west of Yellowstone Creek. The cool morning doesn't seem to bother Celeste, but by mid-afternoon, she is quite tired. They camp at a flat spot near Yellowstone Creek north of Park Creek. It is only twelve miles from where they started, but she is wheezing and sweating and shows little recovery when they rest. He feeds her some high-energy pellets that were for later, when they were above the forest line, hobbles her in the field, and goes to the creek for some fishing.

It has been years since these trout have feared anything but winter ice. Jim simply lowers a line with his state-of-the-art lure and the trout take turns snapping it up. He returns to camp and cooks the fish in a breadbox-sized gas stove while the dung stove keeps the tent warm. He doesn't really need to keep the tent warm yet, but what he doesn't burn, he carries.

Jim naps, and the sun sets. He wakes after dark and checks on Celeste. She is feeling much perkier. The rest and familiar food are doing wonders. He takes time to study the Milky Way. TV viewers get an equivalent view from Kitt Peak Observatory in Arizona. The view from both sites is stunning, but working with the astronomical equipment allows the network to enhance the image considerably. Only a few amateur telescope geeks call in to complain that the stars are shifted too far north for this to be a view from the Uintas.

Chapter 3

The light of day comes long before sunshine in Yellowstone Creek, and at this altitude, there is always a chill in dawn air. As Jim leaves the tent, a herd of antelope is watering at the creek, some of perhaps two hundred thousand in the Uintas by last satellite count.

"This man, this intrepid adventurer, is armed only with his journey recorders and a knife," Chet Bradley reports. "He must be well-versed in the ways of the wilderness before ever arriving. There are bear and wolf here as well as the much more dangerous marauding buffalo."

Jim is picking up the satellite feed of Bradley's comments through his helmet. As he breaks camp, he thinks, *Fifty weeks! How time has flown. Last year's wilderness traveler came out, and they held the final drawing two hours later on national TV. It was so goddamn silly, like Miss America. Regionals, state-wides, and finally the nationals. Hell, somehow, I got it.*

He mounts Celeste, and they ford the Yellowstone. On the far side, they enter a stand of quaking aspen for the first time. At noon, the easy path of abandoned road ends, and a narrow trail takes its place. Jim takes a break. He leaves Celeste in a field and picks his way through the trees down to the creek for some more fishing.

In the fir grove off to his left, Jim spots movement. Jim turns right and sprints to a thick stand of aspen. Never looking back, he scrabbles up the smooth, white bark of one with the aid of the neocollagen climbing claws he pulls from his vest. Behind him comes the pounding of buffalo hooves. From twenty feet up he looks down and then crawls up a bit more. Beneath him is a solitary bull buffalo standing nearly ten feet high at the shoulder. It snorts at him, scars the bark with his horns and hooves, and then circles the tree.

The camera on Jim's helmet catches it all. Tinkerbell missed the initial sprint but got great shots of the bull trying to climb after Jim and circling the tree.

"That was a close one," Jim whispers into his mike. "Did you see how fast he was? I think he came out of a gully hidden by those firs. He was certainly pi … an aggressive one, wasn't he? Certainly had no fear of man, either. Amazing what a difference just a couple generations of pristine living makes."

Half an hour later, Jim is still up the tree, motionless.

"Patience," Olson told him at their first meeting. "To survive in the wilderness, the first virtue is speed, the second is patience, and the third is knowledge. Speed we all applaud, so you have had plenty of practice learning that in your everyday activities. You've also had long experience acquiring knowledge. But civilization abhors patience. It fights learning patience every day with instant gratification. Learning patience will be your biggest challenge."

"How will I learn patience, sir?"

"The same way you learned to acquire speed and knowledge. With practice, long and painful practice." Olson looked at him for a full minute. Jim wanted to fidget in the worst way, but he held still. "Can it be done? Can you learn enough patience in less than a year? We shall see."

"I will learn patience. It sounds like a good thing to have."

"It will certainly be a good thing for your adventure! But be warned: You haven't been taught patience before now for a reason. Civilized people don't like patience. Media people are among the most civilized of people, and one day, they will cut you off because you have it. They are not patient people."

Olson took off his glasses and wiped them carefully. He put them back on. "But you may be pleased with the results of your efforts, even if the media people are not."

The bull moves on. Half an hour later, Jim slips down the aspen and returns to Celeste.

Chapter 4

The rest of the day is a tough one. The trail winds through brush and groves of aspen so tightly spaced that Celeste's wilderness skirt keeps getting stuck. Jim is on and off her, manhandling the skirt through tight spots.

"Control, the problem we've got here is that the droppings are getting into the joints and jamming them. The skirt is getting stiff. I can get by without the dung stove. I'm going to abandon the stove and Celeste's wilderness skirt here."

"Jim, you can't do that," comes the nervous reply. "You have to carry out what you carry in. We supplied you with the dung-burner mostly so you wouldn't have to pack out Celeste's dirty work."

Jim sighed. "Okay. I'll try taking it down to the creek and washing it out."

"Jim, you can't do that either. That creek is part of the watershed."

"What the! ... Fine ... Check the satellite images. Is there a damp spot, a puddle, near me that isn't considered part of the watershed?"

For the next couple hours, Jim freezes his fingers trying to wash out the skirt in a nameless, reed-choked marsh. He makes camp at the base of a pine-covered cliff and broods.

The next day breaks frostily. Jim tries shaking out Celeste's skirt. Instead of being supple, it is a board. The water from washing is frozen in the joints.

"Patience, patience," he mutters. He spends the morning fishing and wandering in the local area looking at plants, birds, small mammals, and rocks. The big find of the morning is some bear tracks. He checks his climbing claws. Finally, after the sun is high in the sky, the air and sunshine defrost the skirt, and once again, Jim and Celeste are underway.

The going remains hard. They cross the ten-thousand-feet line, and Jim walks Celeste more than he rides. The thin air rapidly tires them both. The Yellowstone gorge widens and then closes in on them again. They ride through pine, but not far above them are gray, barren slopes that tower above the tree line: the peaks of the Uintas.

"These peaks are limestone rock," Jim tells the viewers before Bradley can find his notes. "They were laid down on the bottom of a shallow sea. A few million years ago, well after the age of dinosaurs, they started to rise. During the last ice age, they were sculpted by glaciers into the form you see now." Jim pans his view. "The peaks make a network of cliffs and razor-back ridges that tower over the high plateau."

To the left looms the ridge that leads to Stone Mountain. It is somber in the afternoon sun. To the right is a shining ridge that leads to Kings Peak.

The gray cliffs pick up a sandy, iridescent hue as the sun sets in the west sky. This ethereal beauty is enhanced by the sky darkening overhead: thunderheads. Rain soon begins to fall. Jim and Celeste take refuge in a pine grove, but they are careful not to stand under the tallest tree. Lightning soon follows. It is spectacular, but it confines itself to the peaks, and in a few minutes, the rain ends.

The sun is under the clouds, and the cliffs remain spectacularly lit as Jim and Celeste move on to Gem Lake. High up the canyon walls, they see herds of bighorn sheep.

Chapter 5

"It's Day Seven of Wilderness 2050. This is Chet Bradley, and our satellite monitors show us that Jim is now camping at Gem Lake." The viewers see a picture-perfect alpine lake surrounded with alpine meadows and pine groves. It is hard for them to tell, but the pine trees here are a lot shorter than those on Yellowstone Creek. Gem Lake is thirteen thousand feet high. "He's only a day away from his goal: Kings Peak, the highest point in the High Uintas. Before this primitive area was established, Gem Lake was a popular camping and fishing area." Viewers see a still shot from the late 1950s showing rows of neatly-pitched tents. "Let's see how it looks now that humans have been excluded from the area for fifty years. Jim, what are you finding?"

"Well, Chet, as you can see from the cameras, the area is just beautiful. I'm watching a sunset projecting its last pink rays onto Kings Peak. And in fact, if you look carefully, I think those are bighorn sheep on the slope." The camera on Tinkerbell zooms in for a close-up shot of the creatures moving gracefully across a boulder field. "The lake itself is clean, clear, and cold, and there is a beaver colony in the creek nearby."

"Thank you for that report, Jim. We'll sign off and let you get some rest."

Floating in the twilight over Jim's head are the silent motorized gliders. In daytime, they float on the breezes along with the eagles and hawks, disturbing neither. At night, they glide alone. They are not there to help or rescue Jim. Rather, they are looking for illegal intruders: poachers, neighboring farmers, and ranchers pushing their wildlife control into the parks; quick-buck photographers; and adrenaline-rush hikers.

Jim remembers sitting in the ancient classroom with only ninety days to go. The windows were open, and Jim felt the spring breeze flowing in from the fresh-cut lawn outside.

"No one has been brought to court for wilderness intrusion this year, Jim," said Horace Manly, the Park Service liaison briefing him on primitive area conditions, "and no one is likely to be. The gliders take care of the problem quickly and cleanly, right on the spot, and a spot is generally all that remains. This Park Service policy minimizes wilderness intrusion."

"Has anyone given you any flack about this? I guess maybe the environmentalists wouldn't, but what about the ACLU?"

"This policy was instituted about ten years ago to clear up a chronic problem in the Mount Shasta area. There were some dope-growing mountain men there who refused to stay out. They claimed that spiritual powers from the mountain suffused their crop, making it unique. They lost in court; we tried relocating them three times. Finally, we put up the gliders over Shasta. All but one or two found somewhere else to farm."

"And those one or two?"

"Haven't heard from 'em. Their estates have brought suit claiming cruel and unusual punishment and lack of due process and such, but it'll be a while before that works its way through the system. The courts really do read the election returns, so we expect to win. In the meantime, we've instituted gliders over all the national parks, and as I say, the intrusion problem has dropped dramatically."

Chapter 6

"Chet Bradley here. Jim will make his assault on Kings Peak from his Gem Lake camp. He must cross six miles of above-tree line meadow and marsh to get to the base of King's Peak. He must then become an alpinist and negotiate the boulder fields and cliffs that surround this highest peak in the Uintas."

Jim is grim from his days of high-altitude traveling, and his muscles ache from climbing on and off Celeste so much. "I didn't get out as much as I should have," he mumbles as he gets in his sleeping bag. He dreams of wrestling wilderness skirts.

He wakes up in deep darkness, half frozen and with a calf cramp so bad he screams. The dung stove has gone out. It is ten minutes before he can maneuver himself out of the tent and try to walk out the cramp in the frosty night air. The cramp finally fades, but his calf still has a deep ache. He is about to go back in the tent when he notices the moon. It is surrounded with a faint luminescent ring.

"Control, I see a ring around the moon. Does this mean what I think it does?"

It takes a moment for the voice on the other end to wake up. "I'll check," it says.

Jim goes in and tries to relight the dung heater. It seems the dung didn't dry out enough to burn reliably. It starts again, but feebly.

"Jim, bad news," Control comes back. "Satellites show a Pacific storm is coming in stronger and more southerly than forecast. It could hit the Uintas hard."

Jim struggles out of the tent and looks around again. "This could get tricky. I've got a south wind here already."

"It'll increase tonight and tomorrow, Jim."

"Tonight, it will bring clouds. Tomorrow showers in the valleys. At this altitude, it could bring snow. Then the wind will turn, and blow long and strong from the north … bringing cold, heavy snow." Jim's voice starts to slur with fatigue and drowsiness.

"That's right, Jim," finishes Howard Rufkin, staff meteorologist. "And once started, the storm could go on for days, or it could end in hours. The long-range satellite reports are inconclusive on the matter. The jet stream is unstable. If it stays over the Uintas, this is just the first of a series of autumn storms lined up over the Pacific ready to batter the mountains. If the jet stream moves, high pressure will build; the storms will go north or south, and the Uintas weather will clear."

"I've decided. I go now. Tomorrow will be the day." Jim is dog tired. He can feel his patience wearing thin. It is a hasty decision, and that nags him. "I'll be making this part of the journey alone. Celeste will stay here with the tent."

As he assembles the few things he needs for this sprint on the summit, he thinks about Olson.

"Hah!" snorted Olson. "You only think you're leaving class. The dentist's office is a field trip for you." He was right. Lying there while the dentist worked over his smile, he was putting his patience training to good use.

The dentist trip Olson didn't mind, but when Jim left for trips to the plastic surgeon, voice trainer, clothiers, trip planners, equipment planners, and business agents, he would just sigh. Finally, with just a month to go, he pulled Jim aside and said, "These are not patience training; they are distractions—impatience training. These people will not tell you, so I will: What good are all these papers and clothes if you don't survive? You have just thirty days; survival must be your first priority."

Now Jim can see why. His legs are shot. He's eaten two-thirds of his food, and because he is already on enhancers, recovery will be slow. He's pushed too hard. He's been impatient. Now, he is in a vicious circle. He has to push harder to beat a storm that would have been a minor inconvenience if he were a half a mile lower.

Jim sorts his equipment, but he can't keep his eyes open.

"Time?" he grunts.

"Two thirty-five," comes the reply.

"I need sleep. Wake me in two hours." Jim crawls back in once more.

Chapter 7

His eyes open; he's had some sleep, but now it is time to stay awake. He is stiff. He wills his legs to move; they cramp first and then finally respond. He staggers in place, working out the cramps. The moon is high and shining between smooth, pale clouds. He takes his first step toward a nearby bush and then another. Finally, his legs warm enough to move smoothly, but they still ache. He hefts his pack, and the final push begins.

Jim takes the old skyline trail through Tungsten Pass and up to Anderson Pass. For two hours, he walks the moonlit trail through a rock-strewn meadow, circling up and around the north face of King's Peak. Then, as the eastern sky brightens, he leaves the trail and cuts up and over the giant rocks of the talus cliff on the east face. The brisk south wind races up the slope without chill. Jim walks and climbs until he is overheated and panting and then he breaks. When his wind is back and he starts shivering, he starts walking again.

The eastern sky turns blue and then rosy. The boulders Jim is negotiating turn to rocks, then scree, then a rock face. Jim is near the top. He pants as he struggles up the final few feet of the cliff to the summit.

Just as the rosy glow of the sun shining on the bottoms of the clouds races across the sky ahead of the dawn, Jim makes the top. As a symbol, he takes four pictures: north, south, east, and west. His helmet-mounted video and Tinkerbell frame the shots much better than he does, but even Jim's amateurish efforts produce spectacular results. To the north and south, Gilbert Peak and Mount Emmons strain to reach to the pink dawn clouds, which are so close they move and tear visibly as they cross the crests. To the west, sunlight brightens just the crests of Mount Powell and Wilson Peak. They stand out white and brown from the dark gray of unlit clouds above and gray-blue-green of the forest below.

The wind is so strong now that it nearly blows Jim off the crest. He moves to the lee of the ridge but stays near the top. He rests and eats and rests some more. His calves still ache from the night before.

All through the morning, the clouds thicken and the wind grows. Every hour, Jim takes more pictures. They are spectacular, but not nearly so much as the first. At two o'clock, the first brushes of mist touch the mountain top. "Sorry, folks. There'll be no sunset shots this trip." Reluctantly, Jim starts down the slope.

The return to camp is a blur. He is tired, cold, and hungry. He can't move fast enough to keep warm, but he tries to and falls. It is not serious. Fortunately King's Peak shelters him from the worst of the wind until he is back onto the High Line Trail. The rain starts before he reaches camp. When he gets there, Celeste is waiting forlornly with her back to the wind and rain.

He wastes no motion slipping into the tent and downing food and water. It is daylight, so he checks the dung heater and removes the wet fuel. He lights it and falls asleep immediately.

Chapter 8

"Jim. Jim." Control finally wakes him.

"Yes?"

"Jim, it's almost dark. You'd better check your campsite." It is a few minutes before Jim can summon the energy to look out. He hears the wind, and sure enough, when he peels back the tent flap, it is pure white blowing snow outside. The storm has set in.

"Where's Celeste?"

"She's a quarter mile downslope in a pine grove. Tinkerbell's keeping an eye on her."

"That's nice. It's the first useful thing I've seen that piece of equipment do."

"Well, it's not entirely altruistic. The floater doesn't do well in blizzards, and neither does Celeste. They both headed for calmer shelter."

A wind gust blasts snow into Jim's face. He shuts the flap. "Calmer shelter doesn't sound like a bad idea." He bundles up and then braves the storm. Outside, the view is pure white from sky to ground, and the wind howls by his head. "Control, I've got a whiteout here. Can you navigate for me?"

"Be glad to, Jim. Right about ninety degrees and then start walking."

With Control's aid, Jim navigates through the wind and six inches snow to the grove and rounds up Celeste. Reluctantly, she follows him back to the camp where Jim hobbles her while he breaks camp.

Jim speaks as he works. "I've wrestled some tents in my day. And wrestling a conventional tent in forty-mile-an-hour winds makes windsurfing look tame. But this new inward-collapsing design really helps. It keeps the structure low and surface area to a minimum."

Jim hefts the tent onto Celeste's back. As he starts to unhobble Celeste, Control announces, "Jim, there are three food packets and a utility case missing from your inventory." Grimly, Jim grabs the palm-sized inventory detector and starts sweeping the ground. Five minutes later, he requests, "Control, guide me back to those pines. The light's fading up here. I'll find them tomorrow."

Celeste has chosen well. The pines fill a canyon that the wind has trouble reaching into. The snow falls vertically here instead of horizontally, and the trees give the land a ghostly perspective in the twilight. Jim pitches the tent once again, feeds and hobbles Celeste, and climbs inside for supper. He hears the gentle hiss of snowfall on the tent as he quietly eats his evening meal.

"Control, what's the word on the weather?"

"Howard Rufkin here, Jim. The jet stream is moving, but it seems to be pivoting on the Uintas. The weather will continue to be unsettled, but how unsettled and for how long is very hard to say. Snow tonight. Both snow and clear skies are likely tomorrow."

Jim sleeps restlessly. The dung stove runs out of fuel, but the snow on the tent traps his body heat well enough to keep away the terrible cramps. Finally, he opens his eyes to bright daylight. He breaks his way through the snow burying his tent to come out to a picture-perfect winter morning: Bright sun, blue sky, and calm air, and twenty inches of powder snow cover the ground and trees. He watches his breath as he trudges over to feed Celeste. She is looking rested, almost perky, but the wilderness skirt is an ugly mess of snow, dung, pine branches, and sagebrush.

"This is going to be a recovery day," announces Jim as he puts her feed bag on. "And first on this list of recoveries are the items I left behind at the Gem Lake camp."

Jim takes his time moving through the snow. The sun is rapidly heating the air and snow, so Jim isn't cold, and the snow is getting shallower but thicker and wetter as he goes. Gem Lake isn't frozen. It reflects the snow-covered pines and cliffs behind it, making yet another picture-postcard setting.

"Control, I'm here. But I'm not locating anything."

"It's probably the wind. We've done an analysis. Try zigzagging to the northeast for five hundred yards."

"Zigging now." Jim trudges for half a day through sunshine and light snow flurries, zigzagging here and there as Control gives him new suggestions. The sun is once again behind King's Peak before the final food pack is found, ten feet offshore in a pond north of Gem Lake.

Jim is shivering when he gets back to camp. He feeds Celeste and inspects the wilderness skirt once again. The snow is mostly gone, but the mess is worse. Dung is piled on dung, and the skirt is clearly not transforming it into stove fuel.

"Control, we've got a problem here."

"Jim, we've studied the problem while you've been gone. You've got to clean out the dirt and branches and then walk Celeste for at least two miles. The natural motion of Celeste's walking will process the dung so it can dry out in tomorrow's sunshine."

"You want me to do this now?"

"The skirt is overloaded already. Celeste hasn't walked for two days, and you haven't cleaned it for three. It weighs about seventy-five pounds now, and it's not going to get any lighter until the dung dries and you clean it out. If you wait until tomorrow, you're going to have to unload part of the dung and then load it back on after Celeste has walked a while."

"Right. I'll start on it." In the cold shadows of King's Peak, Jim pulls branches from the skirt. The delicate work of clearing the twigs jamming the joints must be done with bare fingers. Three times, Jim stops to warm his stiffened fingers. As the sun sets, he takes Celeste's reins and walks her over the meadow. The moon and stars pop out through the patchy clouds in the east as a spectacular sunset retreats to the west. By moonlight, he returns to camp, feeds Celeste, and turns in.

There is snow during the night, but the next day starts bright once again.

Chapter 9

"Jim, the jet stream is moving south of the Uintas. North winds, colder weather today, and a storm this evening. We don't think it's a good idea for you to stay camped as high as you are. We recommend you go back to the trailhead campsite at Grants Spring where you ran into the buffalo and wait for a weather change there."

"Let me see the weather charts," says Jim.

"Here they are."

Jim studies the display downloaded to his helmet. "Could the jet stream be splitting?"

"It could. This is the right time of year for a split jet stream, but predicting a splitting jet stream is very difficult."

"If the jet stream is splitting, that changes the forecast, doesn't it?"

"It sure does. This coming storm will split, and the Uintas will not receive its full force."

"I see a splitting jet stream in that weather chart."

There was a sigh. "Jim, you're not the expert."

"Not the expert? I've lived next to these mountains for most of my life. My dad lived in these mountains all his life. You are a weather expert. I'm a living-here expert. You confirm for me that in this day and age that you're confident of about what the weather is going to be on this peak for the next twenty-four hours. Don't tell me about what you will be able to predict ten years from now. But tell me that you're confident that you can tell me about tomorrow's weather here."

Another sigh. "Jim."

"I see a jet stream split coming. I'm going over Porcupine Pass today, Red Knob and Dead Horse tomorrow, and Rocky Sea the day after. I'm late for Mirror Lake!" Jim starts breaking camp.

"Jim, you can take the old National Rec trail by Moon Lake. It's sixty-five hundred feet lower."

"No. I know the route you're talking about. I came to see the High Uintas, not the Middle Uintas. And I'm not going to buck that horse's wilderness skirt through another fifty miles of forest."

In ten minutes, Jim is on his way. He walks toward a scatter of rocks that were once a shepherd's high-pasture hut. A bighorn ram standing on one of the walls bleats a warning. Moments later, half a dozen ewes bound out of the ruin. Jim trudges grimly on.

At Tungsten Pass, he turns northwest instead of northeast and crosses the snow-covered cirque headed for Porcupine Pass. Traveling is a bit tricky. Snow covers the still-unfrozen ground, so the marshes are hard to identify and there are stumbles and wet surprises every few minutes. But the snow also smooths the way for the wilderness skirt. Celeste is doing well. Being on the move again puts a spring in her step.

On the far side of the cirque, the trail climbs a cliff wall. The trail is a three-foot wide cut in the side of the cliff. Jim leads Celeste over the rough trail. The wilderness skirt flops awkwardly over the edge, but Celeste compensates nicely. Jim kicks hoof-threatening rocks off the trail. When they are too big for that, he levers them off with his shovel. The going is slow, but Celeste seems to be getting her mountain legs.

The sun has been shining on the cliff side all morning, and the snow is nearly gone. The sun is high when they reach the top, and Jim breaks for a simple lunch. At the top of the pass, the trail skirts the side of a nearly vertical west-facing wall as it descends into the Oweep Creek Valley.

As they rest, Jim idly watches as rocks tumble one by one down the surrounding cliffs. Every few seconds, another starts spontaneously. Some are fist-sized, some are head-sized, and once in a while, an even bigger boulder starts tumbling down the face. They roll and bang along until they settle on the talus slope below the cliff. There, the small ones stop quickly, and the large ones roll on and on, headed for the bottom of the cliff.

As he watches, it occurs to him to say something. "These mountains are still active. Earthquakes still lift them higher, and erosion still wears them down. On these high barren slopes, erosion manifests itself as falling rocks. They come down all the time, but they're particularly common when the sun shines on frozen cliffs and thaws them. I'll be keeping a sharp eye out for them as we descend."

Rested and enjoying the sun, Jim and Celeste start down the trail. Jim watches carefully for falling rocks and sees many, but none come close. They reach the snow-covered portion of the cliff-side trail three hundred feet above the cirque below. The snow makes the rock slicker, but the ice is thawed. Celeste's hooves cut through the snow, and except for a bit of stumbling, she is doing fine.

Jim is 200 yards along the trail when he feels it start.

"Crap."

There is time for only one action. He dives for a niche he had already mentally marked as a rock haven. He hits the niche prone as the wave of slush and stone tumbles over the trail: avalanche.

Avalanche! It is a seconds-long ice shower for Jim. His brain reels at the cold shock. He gasps for air; chokes on a mix of ice, rock, and air; coughs violently; and chokes more; but his body stays put in the niche.

When he can stand it no longer, he lifts his head. Trickles of muddy water fall onto the trail. He is in eight inches of muddy slush. The trail in front and behind him is covered.

Celeste! There is no sign of her.

"Crap," Jim mutters.

"Jim! Jim! Are you there? This is Control."

"I'm here."

"Jim! Jim! Are you there? This is Control."

"Crap," Jim mutters again.

He is freezing, but he doesn't get up yet. He looks around some more. The trickles from above are ending. The slush on the trail settles slowly. Shakily, he gets to his feet and looks around. Ten feet away, the downhill trail continues with just a light snow covering, but for fifty feet behind him, the uphill trail is a mess of slush and rock. He is about to continue downhill when he sees another avalanche cover more of the trail in front of him.

"Crap!"

The whole face is unstable!

He fights his way through the slush, headed back uphill for the pass. It is damn tricky. Sometimes, he is on his legs; others, he is on his hands and knees to keep from sliding off the slush heaped deep on the trail. He makes it to the far side and sprints to the top. There, he falls to his knees, dying for air. He looks back. Behind him, section after section of the snow on the cliff lets loose, headed for the cirque below.

"Crap!" he finally gets his voice to say. "How can just six inches of snow avalanche!" he yells to the sky. "And why just as I'm going down? If I'd started down fifteen minutes later, I'd have seen this and waited. If I'd come an hour later, I would have never known it happened!"

Below, he sees movement. Celeste! It is like a birth from slush. One leg and then two thrash through the slush, and a head breaks into view. Then it ends. She doesn't stand. She doesn't even turn off her back.

Jim waits a full minute. He buries his face in his hands and moans. "She didn't deserve this."

The mourning lasts perhaps thirty seconds before he can't control his shivering.

Off come his clothes. He lays them out to dry while he walks and swings his arms. He warms in the bright sunlight, but puffy cumulus clouds are building at mountain-top level and a haze is once more building high above. The sun won't last long.

The clothes dry quickly. He puts them on. When he puts the helmet back on, he hears, "Jim! Jim! Are you there? This is Control."

"I'm here."

"Thank goodness! You blipped out completely. No voice, no video, no Tinkerbell. What happened?"

"No Tinkerbell?"

"No Tinkerbell. What happened?"

"Avalanche. Celeste is at the bottom. I'm at the top."

"Are you all right? Is she all right?"

"I'm *okay*."

"We monitor you as mildly hypothermic."

"Celeste has a broken back."

"What?"

"See for yourself." Jim watches the horse struggle feebly in the slush. The front of her is upright.

"Oh, this is bad, Jim. We've got the analysts working on it now. Have you got any ideas on how to get her out?"

"What?"

"You've got to pack all your stuff out."

"I'm stranded above thirteen thousand feet with a storm coming, and you're worried about how I'm going to pack out a dying horse? Get real, man!"

Jim stands. "But there's equipment down there I need to survive. I'm going down."

The slope has shed its snow; the trail is once again just a potentially dangerous mountain trail hacked out of the cliff face. Jim descends quickly to the talus slope and then cuts across the acres of slush to Celeste.

Celeste sees him coming and whinnies plaintively. She struggles once more to rise, but only her neck and front legs are paying any attention to what her brain is requesting.

Jim pushes her inert rump around until he can extract her feed bag from underneath her. He puts it on her, she starts feeding, and Jim briefly strokes her forelock and mane. Then he drives his knife deep into her neck.

"Jim, what are you doing!" comes from Control. He works the knife until he is sure an artery is cut.

One sob of grief is all he allows himself as he pulls the knife out and cleans it on the slush.

Chapter 10

With one eye on the gathering clouds and the other on his object detector, Jim searches the slush field for his equipment. Jim finds his tent, stove, and food packets. Some of this stuff is on the surface, some is deeper. If it is deep and not useful in his current situation, Jim leaves it.

Tinkerbell! There it is on its side, buried about six inches deep. Jim walks on.

"Jim, you can't do that."

"I just did."

"You're in there to serve the public."

"They've been served."

"Tinkerbell can help you."

"How?" Jim finds another food pack and rescues it.

The hiss of the communications line increases slightly.

"Jim, I'm talking with you on a private channel." It is Olsen, "Please don't acknowledge that you hear me." Jim keeps searching. "You aren't the only one upset by this accident. The animal-rights geeks have flooded the Park Service switchboards and Richard Moonan, the park chief, says he's got a leak from one of its underground bulletin boards about a rock-throwing demonstration somewhere this evening. He's ready to pull your plug."

Jim stiffens a little and looks up and down, up and down, in a sort of a half-nod.

"One of my friends in the service heard him wondering out loud how likely a beeper failure would be at this point, and some rooky brownnoser was promising him a full report on failure modes within the hour.

"Jim, the Park Service doesn't control Tinkerbell. It's a media thing, and it's another set of eyes that you'll need if the Park Service turns on you."

Jim looks up and down again. "I was just thinking of the bighorn sheep I see up here. I know what I see in them, but I wonder, what do they see in me?"

"Ratings, Jim. Right now, it's you against the wilderness. If the Park Service acts, it's you vs the wilderness and the Park Service. The ratings will skyrocket."

Jim's searching brings him back near Tinkerbell. He looks at it and gives it a vicious kick.

"Silicon-for-brains piece of scrap!" he spits out and then walks back to Celeste. Tinkerbell isn't completely free, but it is lying upright and the sun will finish freeing it in about ten minutes.

"Thanks," says Olson, and the private line cuts off.

There is a wide red stain around Celeste's neck. Jim sits on her and does a final sorting out of what he will carry.

Finally, he stands and faces her. "Celeste, you weren't much of a mountain horse, but you deserved better. You had the heart, just no head for this kind of stuff." He kicks the skirt. "No horse should have to wear such a claptrap."

He bends down and unfastens the side he can reach; the other fastener is hopelessly buried. He removes the beeper, yanks the skirt, and twists it around until it stretches over her.

"It wasn't much of a skirt, it isn't much of a shroud, but it's the best I've got. Goodbye."

Under a gray sky filling with rain or perhaps snow, Jim walks out of the treeless cirque down Oweep Creek.

<<<*>>>

He is moving quickly. Two miles away, he is passing the first scrubby pines when he hears a series of explosions echo off the cliff walls.

Behind him, a Park Service glider is vaporizing the equipment he left behind as well as Celeste. The slush vaporizes, creating explosions and a white cloud of mist. He continues walking until he notices the glider is moving toward him and the shooting isn't stopping!

"Control, what's going on? Why's that glider still shooting?" He runs down the slope beside the creek.

"We're checking, Jim."

The glider keeps coming. It is no longer over slush, so the shots aren't as clearly marked, but it is clearly chasing something and that something is dodging. The glider circles back a couple times. Jim is out of breath. He slows to a trot. The ground is rough here. He is gasping for air.

"Control!"

"We've got no explanation. We're trying to contact the Park Service now."

The glider is banking back toward him, and it is only one thousand feet behind. No turning away this time, and no outrunning it. Next to the creek, he spots a rough overhang between two rocks and races for it. Ten seconds before the glider arrives, he dives in, turns, and looks out. The glider is lined up on the creek. He hears a snap each time it shoots. He sees the target: Tinkerbell!

Chapter 11

Tinkerbell dodges and gyrates wildly as it follows the creek bed headed for Jim. It hides behind trees and zigzags across open ground. It is poetry in motion, and the glider-targeting system is not up to the task of zeroing in on the floater. Jim smiles at the sight. When Tinkerbell gets sixty feet from Jim, the glider loses interest. It passes overhead, rises lazily on the air currents once again, and soon disappears into the overcast skies.

"Jim, the target is Tinkerbell. It doesn't have its own beeper, so when you left, it became eligible. The media guys got it pulled out of the slush just as the glider came up, and they've been playing Astro Fox trying to get Tinkerbell close to you. The Park Service says it can't do a thing, but you're in no danger as long as you've got your beeper."

"Thanks for the update," sighs Jim. His breath is back, and he continues down next to the creek bed. Behind him, the clouds coalesce and the peaks disappear in a rainy, snowy mist. The cold north wind catches up with him and the wet mist soon follows.

"Jim, you've got to keep moving; there's no split. The jet stream has moved south, and the storm will be strengthening."

There never was a trail down Oweep Creek. Sometimes, Jim finds a game run, but most of the time, he is picking his way down a steep pine-covered slope. His knees start aching, and he starts stumbling.

He rests, but while he rests, the rain gets stronger.

It is late afternoon. The west side of Oweep Creek Canyon ends, and Jim breaks onto an old north-south trail west of Lake Fork River.

"Jim, Moon Lake is four miles downstream, and there's an old campsite two miles beyond that on the west side." Jim looks around. The trees are tall enough here.

"Time for a break." He climbs under some pine boughs, sits down with his back leaning against the trunk, pulls his knees up to his chest, and spreads his poncho over his aching knees. He pulls out a food pack, and after admiring it like a fine wine, he sensuously squeezes out a meal. Jim knows he should push for Moon Lake, but the shelter, the meal, and the memories of the morning overwhelm his good sense.

Jim sleeps.

Chapter 12

Jim wakes to darkness and a wet bottom. It is raining hard; there is a flash of lightening followed by a clap of thunder. He wills his aching knees to lift him and bumps his head on a pine bough. He grabs it and continues to rise. The tree is soaked, as is the ground. It is dark.

"Control, what time is it?"

No answer.

"Control?" He pauses longer this time. "Just like at the avalanche. This helmet must be failing when it gets too wet."

Jim struggles out from under the pine. It is not yet pitch black, but the dark storm clouds and high cliffs have created a dim twilight that will soon fade to black. Jim starts briskly down the trail, ignoring the lightning flashes and thunder rolling about in the canyon.

The trail is full of water, and the rain is coming down hard when he rounds a bend and comes to a steeply descending section. He is using the lightning to guide him, so his head is up when the flash comes. Below and stuck in a tree in the flooding creek is the body of a deer. Two wolves are splashing in the water trying to reach it. Three are sitting on the edge watching.

Jim watches them and considers. He should be able to sneak by the wolves. The storm will mask his sound, and the deer meal will satisfy them. Jim starts down, but his footing betrays him and he slides most of the way. It is not pretty, and it is not graceful, but he arrives at the bottom with injuries only to his pride—and an audience of three wolves.

They watch Jim rise and walk away from them uninjured. They don't follow; there is an easy meal in the stream to attend to.

Jim moves on but not far. It is just too dark. He climbs about twenty feet above the trail, paces off a spot in the pines where water is mostly going around it, and pitches his tent. It is cold and damp inside but not as cold and damp as outside. He shivers violently as he places his shirt and pants where they will dry and then climbs in the sleeping bag. It takes a while, but his body heat finally spreads through his limbs. The shivering subsides.

<<<*>>>

It is still dark when he wakes again. He hears the wolves howling. The rain has stopped. Jim feels rested. He looks out. The sky is crystal clear with a quarter moon riding high. It shines on two inches of snow, and visibility is much better than it was during the storm twilight. Jim is cold but not deathly cold like he was in the storm. His clothes are still cold and damp, and it is like diving into forty-degree water putting them on, but the material doesn't hold much water, so the discomfort lasts only a few minutes. He puts on his helmet.

"Control?"

He folds the tent and packs it.

"Control?"

"Is that you, Jim?"

"It's me. This helmet is cutting out in wet weather."

"Good to have you back in contact. How's it going?"

"All right. I saw some wolves earlier."

"We saw them too with Tinkerbell. In fact, they're still around. Tink spotted them snooping near the tent about an hour earlier."

"What time is it?"

"Four-fifteen. You're starting early."

"I'm rested. The light's good. It's time to make tracks."

Jim breaks camp, climbs back down to the trail, and follows it. Soon, he reaches Moon Lake and follows the trail down the west side. Moon Lake would be considered a small- to modest-sized lake in a place like Minnesota, but it is a giant in the Uintas. It fills the canyon it is in almost from cliff wall to cliff wall, and forest surrounds it. As Jim walks the trail, the moon's reflection sparkles at him in the lake. Once again, he's struck with the stark beauty of the wilderness.

At the south end of the lake, Jim crosses a wide meadow filled with young aspen. The lake shore is easy to reach, and there is a touch of blue in the sky. Jim stops to fish for breakfast. He is putting his line in for the third time when the bear comes lumbering out of the woods.

"I guess this is still a popular fishing spot," muses Jim. He retreats up the hillside to prepare his catch and watch the bear work the same hole. It is a sight! Half the bear's technique seems to be splashing so much water out of the hole that the fish have nowhere to swim.

The sky brightens. A second bear shows up. The two bawl at each other a while before they each settle in their spot.

"Jim, off to your left."

The wolves are watching him from the other side of the meadow. Jim keeps eating.

"Send Tinkerbell."

As the floater moves their way, they move into the woods. A glider appears, skimming over the west ridge line. Tinkerbell beats a hasty retreat back to Jim.

"The Park Service has you on a tight leash, Jim. The Zero Population Group is filing suit, accusing the Park Service of negligence because of Celeste. Claims you doubled the amount of trash left in the park this year and that it was irresponsible to let you bring a horse in."

"As I recall, they've been waiting for something like this."

"Uh-huh, and the People Second group staged a rally at the Lincoln Memorial. And there was a skit where a person gets a stroke and the horse cuts the person's throat. The horse says to the person, 'If you can't think, what good are you?' It's getting ugly out here."

"Political entertainment at its best."

"Truly, but I don't think Richard Moonan's enjoying the show."

Jim finishes breakfast. Giving the bears wide berth, he heads south.

"Jim, the old National Rec trail heads west from where you are."

"I'm not taking that. I'm coming out the way I came in."

"This is a switch. Care to explain?"

"I've visited Kings Peak, and Kamas is an awfully long trip from here without a horse. I'm headed out."

"I didn't think you liked traveling with Celeste that much."

"I didn't, but things have changed. There's one more thing I need to do, and I'm out of here. Do me a favor, will you? Guide me to a set of coordinates. Put me right on the money. Okay?

"The coordinates?"

"110 degrees, 22 minutes, 32 seconds west; 40 degrees, 23 minutes, 27 seconds north."

"Mighty precise."

"Put me there on the money, okay?"

"Okay."

Chapter 13

Jim hurries south. Clouds build again, and a cold wind pushes him on. Fifteen miles from the gate, the snow catches him. This time, it starts thick, cold, and wet. Twelve miles from the gate, Jim is trudging through inches of slippery snow and blinding wind. He is carrying two walking sticks. He stops.

"Which way?"

"West about a quarter mile."

The hum of the private line comes alive.

"Jim, this is Olson. Where are you going?"

"It's personal."

"Don't talk to me, Jim. The Park Service doesn't know about this line. Word is going up the chain that you're getting erratic. Moonan doesn't need to hear that. Is there anything I can pass up the line to help make sense of this?"

Jim trudges on silently.

"Very well." The private line hum cuts out.

The snow is so thick that Tinkerbell is having trouble tracking. Twice, it almost brains Jim and then flies off drunkenly. Jim takes a swing at Tinkerbell with one of his walking sticks.

"Back it off, guys. If Tink can't fly in this, neither can gliders."

"What? You're breaking up, Jim."

"Crap, the helmet is getting wet." Jim takes it off and wipes off the snow as best he can. He puts it under his coat.

"I've seen stupid, and I've seen stupid," he mumbles, "but protecting my head protector from the elements has to take the cake."

He puts it on. "Control? Where to now?"

"You're back! Let me check. Two hundred seventy feet bearing 030 from your current location. Plus or minus fifteen feet."

"Got it."

In spite of his profound exhaustion, Jim trots the last few meters. He is in a broad grassy field with nothing to distinguish this spot from any other.

"Here?"

"Best as I can tell. You're there."

Jim hammers one of the walking sticks deep into the ground. He uses fishing line to tie the second one to it as a crossbar. As he works, he talks between gasps.

"This town … had a cemetery. In that … cemetery … at this very spot, two people were buried. My mother … and … father. It … isn't proper that … they rest without a mark."

He finishes.

Jim looks up at Tinkerbell. "*Screw you*, Park Service!"

The private line comes on again. "Jim, now is the time for speed. Head for the gate and head there fast."

Chapter 14

With a maniacal giggle, Jim trots to the southeast, heading for the road out. But he doesn't trot long. Lean, gaunt, burned from days of exposure and strenuous exertion on short rations, Jim quickly slows to a fast walk. Light is fading. He stumbles into a ditch and falls. He hears the unnatural snap and feels the pain.

"Damn. I broke something."

There is no response.

"Ankle or something."

He begins a nightmarish limp out.

"Patience," he whispers. "I must have patience."

The snow and wind continue.

In the near dark, he spots a visitor: One of the wolves. Jim throws a snowball at him, and the wolf backs off, but he's seen Jim limp. He follows.

About half a mile from the gate, the snow is a foot deep, and there are three wolves. Jim can see the gate— at least sometimes. He can hear the cheers, and they call him onward. He is close, but only he can negotiate the distance. Only one person is allowed in the park.

The wolves lose patience. One leaps at Jim and clamps on to his good ankle. Jim goes down. A second goes for his throat. There is a blur of motion, and a howl of maddened pain. The wolf at Jim's throat thrashes and backs away, leaving a trail of red behind. In Jim's hand is a climbing claw, with a foot-square patch of wolf hide jammed in the teeth. Tinkerbell shows the wounded animal thrashing in the snow, trying to lick the pain from the exposed ribs on its left side. Jim faces the third animal; it backs away. Jim looks at his feet; the claw moves up; and the first wolf releases his ankle before the blow can fall. The two healthy wolves wait once again for two meals, Jim and their pack mate.

"Jim, Jim! I don't know if you can hear me, but your beepers have been deactivated. Deactivated, Jim! They will no longer protect you! You've got to get out before a glider finds you! Jim, do you hear me?"

"I hear you."

"Jim, do you hear me?"

Jim waves at Tinkerbell.

"Good. This storm is covering you, but it's going to end shortly. Jim, you've got to move quickly!"

Jim moves, but not fast.

With a third of a mile to go, the snow stops.

After a bit more, Jim sees a star.

"It's not happening," he says through gritted teeth.

"Jim, we can see you. Please hurry."

No answer.

"Jim, what are you doing now?"

"I'm taking off my clothes. Damn, it's cold here."

In two minutes, Jim is naked to the world. He throws his clothes and equipment every which way, and he continues limping on.

Two minutes later, there are a series of explosions with clouds of mist behind him.

One hundred fifty feet to go.

Jim freezes. The glider flies almost invisibly overhead.

The glider turns back to finish off the rest of Jim's equipment. Jim moves another hundred feet, and again, he freezes as another glider approaches.

Jim remembers clips of helicopters in Vietnam. One comes in while the other leaves. The Vietcong had to stay put for hours at a time.

But the approaching glider veers and starts shooting off to Jim's left. It is shooting at a moving target. Tinkerbell!

Jim moves another thirty feet. The second glider goes after Tinkerbell.

Jim reaches the gate and beats upon it three times. With tremendous cheering, the gate opens and a flood of light falls upon Jim. Surrounded with doctors and media people, smiling and waving feebly, he is whisked from the door to a specially prepared limousine-ambulance.

Chapter 15

For the next two hours, the media fills in with canned prejourney interviews and recaps. Then come reports from the doctors and finally the climactic hospital interview.

Once again, Jim is smiling broadly.

"That was a hairbreadth ending, Jim. And it certainly left the Park Service with a surprise. What were you thinking those final moments?"

"Well, Chet, I was remembering my childhood. I was remembering when I used to ride horses up and down the streets of Mountain Home and in and out of the surrounding hills. Those were fine times for me and my family. How was I feeling? I was feeling good. I'd paid my respects to my parents and their heritage."

"So what are your feelings toward the Primitive program?"

"Well, Chet, it's a program with noble aspirations, but it has meant a lot of hardships." Jim sees Olson unobtrusively shaking his head.

"What's this bullshit I'm spouting?" He is ranting now, and Chet loves it, even with the expectable bleeps. "Hell, Chet, I don't know what I was thinking back there. My mind was as stiff as my body. I didn't even know what I'd done until I saw the tapes. I don't remember a thing."

"Doctors report that temporary amnesia is quite common in exposure victims. I'm sure Jim will be much more coherent after he gets a night's rest. This is Chet Bradley. Good night."

Chet leaves. Olson comes up to Jim's bedside. "You made your point, Jim, out in the wilderness on that field of snow. To speak it again here, in the civilized world, will be much more costly to you and those who believe as you do. As things stand, you've got a substantial reward coming for your labors. If you feel strongly about the wilderness program, secure the reward and use it to further your beliefs. Goodbye for now, Jim, and congratulations. You learned enough patience to survive the wilderness. Now you must relearn a different kind to survive civilization again."

By morning, Jim is much recovered. He is interviewed by Chet again as well as other anchormen. They are careful to throw him softball questions, and he is careful to answer with heroic modesty. Once again, Jim is in the world where straight teeth are more important than straight talk and stamina is something you show at the negotiating table. When the media giants finish, the second wave breaks into the hospital room: the agents and lawyers. The media rights to this story will be worth millions.

"Patience," Jim thinks as he smiles. "Patience."

Putting Human Consciousness in Cyber Space

Just as man has long aspired to soar like a bird, he has long aspired to move his conscious out of his body and into other realms. The 2050s version of that will be moving from the human body into cyberspace and back into a different organic-based body from the one it came from.

This is an exciting aspiration, but it will be difficult to achieve, much more so than flying. In this essay, I will outline why.

The Thinking Stack

The heart of the problem is that consciousness is an emergent property of nerve activity. It was not the goal of developing a nervous system, it appeared (emerged) simply as a way to organize analytic thinking processes. For this reason it is quite elusive, as in, hard to describe and analyze. Science has yet to find a "seat of consciousness" in the brain. Instead it comes about when many sorts of neurological activities happening in many parts of the brain coordinate in subtle ways.

Consciousness is at the top of the human thinking stack and emergent. This means that it is very sensitive to the design of the thinking stack. This means that if consciousness is going to have a comfortable home in cyberspace, the whole human thinking stack must be emulated, not just some tip of the iceberg activities that are a seat. This is going to be a big task. Cyber intelligence and self-awareness will have its own thinking stack, and it will be completely different. It will be based on integrated circuits, bandwidth, and data communications standards such as 802.11 and OSI, not neurons, synapses, and neurotransmitters.

What will it be like?

This apples-and-oranges difference means that putting a human consciousness in cyberspace will take enormous resource. And it will run dog-slow compared to its self-aware cyber counterparts. This means that the story line of a human consciousness taking over cyberspace is as fantastic as Santa Claus existing. It won't happen. Instead, the cyber intelligences will be nurse-maiding the feeble human consciousness while it exists in cyberspace.

Given this constraint, the basic purpose of putting human consciousness into cyber is to store it until it can be put back into something organic. Once back in organic, it can mesh much more easily with the nerve-based thinking stack the organic will have.

This still leaves open many interesting possibilities. For instance, the consciousness can be transmitted from one star system to another and inserted into an explorer organism grown on the distant star system. This may also be a solution to the conundrum of youth being wasted on the young. A wise old mind can be inserted into a youthful body.

The process of inserting is likely to be a gradual one. As the new organic grows from immature to mature, the external consciousness will be steadily added. This can happen as the organic is experiencing growing up in a humanlike ways or while it matures essentially motionless in a vat.

In sum, there are lots of interesting possibilities, but having a human mind controlling things while in cyberspace is among the most expensive and therefore the least interesting.

Youth is Wasted on the Young ... Hah!

Moving the human consciousness into cyber space is the modern equivalent of aspiring to soar through the air like a bird. As with flying, it is not going to be as simple or easy as humans have wished for, and there will be surprises. One of the big ones is that humans are unlikely to control much while they are cyber entities. Their thinking will be too slow and clumsy.

As such, humans will not be able to control much while their consciousness is in cyber space. But when we develop the technology to move a human consciousness back into an organic being—an android of some sort—then the thinking can get fast again. And when we have this ability to go both ways, the transfer of consciousness between human and cyber space will make a difference in how people live in 2050.

This story is an example of how.

Cast and Facts

Jimmy-863: First consciousness in the young body

Andy-098: The instructor muse for Jimmy-863

Oscar Wilson: An old man put in Jimmy's body

Aaron: New android base commander on Europa

Dave-879: The cyber commander who has been on Europa a year

Dianne, Franco, and Patsy: Three other newly arrived old mind humans

Europa: Orbiting Jupiter, five astronomical units from earth (eight astronomical units is a light hour)

The Awakening

I'm Jimmy-863. Well, I am for now. But I'm growing up. And as I do, I'm learning lots of new stuff, some of it in a strange way. Well, according to my teacher, Andy-098, it is strange.

I play and learn each day and then I sleep. What is strange is that while I sleep, I learn things. I'm learning about things I haven't experienced. It's kind of like having a vision, but I have them every day. And unlike regular dreams, I remember these! I remember them real well.

In school, I started learning arithmetic last week. Then I started dreaming about it. This week, I know how to multiply up to one hundred. Andy says I'm learning real fast, so I will need a new lesson plan next week.

And today, when I came in for breakfast, the first thing he said to me was, "Who are you today?"

It was a strange question. "I'm Jimmy-863, just like yesterday." I replied and started eating. I thought about it and then I asked, "Why did you ask me that?"

"Because one of these days, you'll wake up and be someone new. I just want to know when it happens."

"That sounds strange," I said and kept eating.

"It would be for most people. But for you, it is your destiny. You are growing up to be the receptacle for a consciousness."

"A consciousness ... you mean what I'm thinking?"

"Very much so! My goodness, you're such a bright child already. This is going to be so fun. It's just like when you get older, you will be taller and stronger. But this change will happen sooner. It should happen any day now."

"Yeah. If you say so," I said, and I left it at that. There was a full day ahead of me, and there was much to be learned.

<<<*>>>

That night, it happened. When I woke up, there was another me in me. He was Oscar Wilson. I just knew that. He didn't have to talk to me or anything. I also knew he was raised in Shaker Heights, and his dog was Harvey, but there was a whole lot I didn't know.

"Who are you today?" asked Andy at breakfast.

"I'm Oscar Wilson," I answered proudly.

"Ah, good. The change has started," he replied.

"But why am I floating in a tank? Why am I wearing a breathing apparatus? Why does my body look and feel so strange?" asked Oscar.

"Because your body is designed to live in low G. When you grow a little more and finish transferring your consciousness, and we know this process is working. Your consciousness will be sent to a space colony being established on Europa, orbiting Jupiter."

"Wow! This is big bunch of surprises for an old man. But, then again, actually completing a consciousness transfer is a big surprise. This was just a wild-eyed experiment when I moved out of my near-death human body ... how long ago?"

"Ten years ago. Things have been changing quickly, Oscar. This colony on Europa is transitioning from totally creation run to creation plus android. It is going to get a lot bigger. We have discovered how to make Pencrock there, and that's going to make a bunch of billionaires.

"There will be lots of surprises in making this transition happen. This is why we are sending you, a wise old man, to be a pioneer in getting it established. You're a wise old man in a young body.

"In your case, youth is not going to be wasted on the young."

"In fact, you're not going to be in this body long. This one is another experimental one. There have been several already. If this one works out well, we will grow one like it, but a little better, on Europa, and have you inhabit that."

"Wow! This moving consciousness can happen a lot."

"That's the virtue of cyber storage."

Oscar spends about two months in this body. It does work out well. It is recycled, and this version of his consciousness is lost.

On Europa

Two years later, its replacement is started in a small-but-growing base on Europa, and seven years after that, Oscar now inhabits that one. When he wakes, he is in his element, and barring accidents and other deadly surprises, he is likely to be in this version of his body for a year to five years. The radiation on Europa's surface is ferocious compared to Earth's, and that sharply limits how long Earth-style life chemistry can survive. When this one fails, he will get a new one, but unlike with the experimental bodies he inhabited on Earth, he will remember what has happened in this one.

Europa Colony Alpha is still a small place, but it is bustling. Oscar is one of ten old consciousnesses in new bodies now moving around the colony. In addition to them are a hundred mechanical robots inhabited by cyber intelligences and ten androids also inhabited with cyber intelligence.

Europa is about a half light hour from Earth, so sending a question and getting an answer from Earth's vast store of information is an hour-long process. Neither the cyber nor the humans on Europa like that—it is strange and unnatural for any Earth-acclimated intelligence—but it is something they learn to live with.

The colony is here, now, and being expanded because the usefulness of Pencrock—a trace compound found in Europa's seas—has been discovered. Getting it to Earth can now be a profit-making activity, but doing so is something that simple automated machinery can't accomplish. Cleverness, human and cyber, is needed on site to make this project happen.

Oscar is now attending his first progress report meeting in one of the base meeting rooms. He is joined by Aaron, the new android base commander; Dave-879, the cyber commander who has been here a year; and Dianne, Franco and Patsy, three of the other newly arrived old mind humans.

Aaron says, "You have all been briefed, but we have been learning lots quickly. It looks like the Pencrock is coming from one of the organisms that is growing in the Europa seas. This means we may be in for some farming adventures."

Oscar asks, "Do we know which organism yet?"

"Not yet. But we are narrowing it down quickly. Pencrock comes out of only one sea, so that's narrowing the choices."

"Is it free-floating? Is it uniformly dispersed? If so, it's likely to be excreted from the organism in some fashion."

"Good point. We can modify our search parameters with that in mind. Thanks."

Aaron looks at Dave. Dave says, "I'm on it."

Dianne goes zombie as she looks over the proposal and then says, "You are getting to the sea through a tunnel, right?"

Aaron says, "That is right."

Dianne says, "If this originating organism is bacterial, which is most likely, you may want to research building vats up top. This may make harvesting a lot easier."

Aaron says, "Good point. We'll keep that in mind."

Dianne continues, "Plus, you'll be able to control nutrient flow and parasites a lot more easily. If this is an ecosystem under the ice, there will be parasites."

Franco now takes his turn. "This tunnel design … it is essentially single wall tube, right?"

Aaron says, "Yes."

Franco says, "This is a place with active plate tectonics. I'd recommend putting in a lot more redundancy. There are likely to be quakes and ice shifting over the years. You may want to double wall and add some melting capability."

Aaron is getting impressed. These old gray beards seemed to have assessed what is going on here very quickly.

Patsy raises her hand. Aaron looks at her.

She says, "I gotta go. And it's time for recess."

He grins and says, "Everyone, this has been good advice. Keep up your research and let's definitely meet again tomorrow. You could be saving this project millions!"

The meeting breaks up, and the kids run out to play.

The First Starship

Mankind is not likely to initiate the solar system's Interstellar Age. It is likely we will first be visited by an interstellar spaceship coming to the solar system to trade. Here are two axioms of this visit.

- This ship will be powered with a constant acceleration, slower than light speed propulsion system. This is a propulsion system that today's physics can support, no wishing and hoping needed.
- It is coming to trade, not extend some kind of galactic empire. And what it trades for will not be anything as simple as gold, silver, or water. It will come for interesting things that are much more complex—much more manufactured and cultural—a Big Mac is going to be much more interesting to these aliens than bars of gold.

This is a story of how such a visit is likely to evolve.

Note that each chapter is split into two parts: an In The Trenches part that talks about the personal experiences of people living during this time and a Big Picture part that talks about the bigger historical and sociological and trends.

Timeline, Useful Information, and Cast

Timeline

2059: First sighting
2062: Foursmoots predicts starship
2074: Starship ETA at Earth
2079: The year-five crisis
2084: (Year 10, new calendar) the Mirondian ships depart

Useful Information

Astronomical Distances

Eight astronomical units (AU) = One light hour
Saturn's orbit = Ten AU
Neptune's orbit = Thirty AU

Gliese 581, red dwarf star, twenty light years away

New Inventions

Spacenet: The solar system-wide communications net

The Main Cast

Anthony and Clairissa: Astronomers who first discover the approaching star ship
John Foursmoots: Astronomy grad student who becomes an anti-Mirondian philosopher
Myrtle Opera: Talk show host
Mirondians: Alien people, all of whom have names that begin with "Miro"
Miro-Sam, Miro-Alex, and Miro-Bob: Mirondian negotiators
Anna Hatchet, Paula Stove, and Mickey Palusi: Solar system negotiators
Jason Furlong, Zeb Noches, and Judy Alonzo: Workers on Earth affected by the Mirondian Revolution
Miro-Robert: Mirondian factory manager and spokesperson
Charles "Good Time Charley" Monson: President of the Moon Boosters

Chapter 1: Contact

In The Trenches

It is the year 2059. Anthony and Clarissa are astronomers. They are working late at night in one of the observatory labs high in the Andes Mountains. (They don't have to be working late at night, or in the mountains, but doing so is part of why they like astronomy.) They are working through the data that were gathered the night before when a cyber system points out to them an interesting anomaly.

They look at the data on the screen. It is strange data, indeed!

"This is just nuts, Clarissa. What are we looking at here?" says Anthony.

Clarissa looks for a while before she answers. "It looks for all the world like a gamma ray star, but gamma ray stars don't exist—not even in theory."

Anthony says, "Yeah, but it is dim in infrared, blue in visible, brighter in UV, and even brighter in x-ray. What can you call it but a gamma ray star?"

Clarissa does some more checking of data. She looks through the historical record of that patch of sky. "And there was nothing there two years ago. It was there, but much fainter, six months ago. Strange ... most strange."

They publish a report and call it GO-1, for Gamma Object 1.

For the lay sky observers around the solar system, this anomaly was nothing to notice because it was too faint to see. For another three years, it remained a curiosity—a small mystery of the big sky.

Then, five years later, in 2064, came a breakthrough.

John Foursmoots, an intensely studious astronomy graduate student in his late twenties, gives a presentation on this anomaly. He is standing at a futuristic lectern and presenting both in person and on the web.

He says, "In summary, this radiation spectrum matches that of what would be expected to be coming from the exhaust of a starship engine, an engine that was headed for our system at near light speed, and decelerating at about one gravity.

"I have on these screens my predictions of how the spectrum will change in frequency range and brightness as the ship continues its approach to our solar system."

This is an astounding hypothesis for such a small anomaly. But it is an exciting one, and it matches the data, so several other observatories now swing to look at this anomaly to confirm data, see if there is anything that can disprove the hypothesis, and see if Foursmoots' projections for its change with time are followed.

At the same time, John becomes an interesting item for science talk shows.

The data follow Foursmoots' projections—the star is both getting brighter and reddening in its frequency spread—and excitement about it grows and spreads. It becomes visible, first to observers with amateur equipment and then to the naked eye.

When this happens, John himself becomes more visible to a now curious public. He makes an appearance on a popular afternoon talk show hosted by Myrtle Opera. John has risen to the occasion. He is a lot slicker now, and his presentation has followed suit. It has a lot of animation now instead of just static charts and graphs.

Myrtle says, "What can you tell us about this starship?"

John says, "We can tell that the engine is powered by nuclear fusion and that the spaceship is a constant acceleration craft, which is currently decelerating. This means that it is headed for our solar system and planning on stopping here. We can also tell that it is ten light years away from us and that it will arrive in eleven years."

"You learned all this from simply looking at a dim star? Amazing! Do we know what the aliens will be like?"

"Well, since we can't build this kind of ship yet, we know they are more technologically advanced than we are. However, we can recognize that it is a ship. It doesn't look like some kind of god magic to us, so they are only modestly more advanced than we are.

"I have talked with some of the technologists at our university, and they estimate it should be possible for us to build this kind of ship in large quantities, the quantities required for space commerce, in, say, three hundred years. If these aliens don't have other surprises in store for us, they should be one to five hundred years more advanced than we are.

"Considering that there has been life on our planet for 3 billion years, and we have been the species we are today for one hundred thousand years, these visitors are practically twins."

"Will they be peaceful when they come?"

"At this stage, all we can know is their starship engine technology. We have no way of knowing their other technologies or intentions."

Myrtle starts talking to the camera, not John, "This is amazing! What a special time to be alive on Earth. We could be witnessing a beginning … ending … and probably a mix both. With that in mind, we have a message from our sponsors."

There is excitement all over Earth and the solar system at this news. Here are some examples:

•   The chattering classes are texting about this all over:

"Did you hear aliens are coming?"

"Yeah. I wonder what they're coming for?"

"Who knows? Could it be women?"

"You mean like Carrie Anne? In your history class? ROFL. They see her. They fly away fast!"

"LMAO."

•   There are the religious points of view:

Preacher A says, "My friends, Jesus is on that ship. When it arrives, Judgment Day will be upon us."

Preacher B says, "Allah Akbar. Allah has sent this ship to show all the people of the solar system the one right way."

Preacher C says, "The Anti-Christ is on that ship. He will be tempting us. Those who succumb will be led astray. The message that ship brings is repent while you still can."

•   The military has its points of view:

Colonel A says, "In sum, we are facing a *War of the Worlds* situation. These alien invaders have technology that is at least one hundred years beyond what we have. We are the native islanders with spears and canoes defending ourselves from an ocean-going frigate with cannons and muskets."

Colonel B says, "Our best hope is to duck and cover. We need to be researching ways to hide our most valuable assets. By valuable, I mean those things we will need to recover our civilization when this ship moves on." He looks around before he says this next part, "This includes our best womenfolk."

Colonel C says, "There are people who argue our situation is hopeless. But before we take that as a given, consider that there are 10 billion people in the solar system. *Ten billion.* If the invaders choose to set foot on Earth, we should throw bodies at them until the last man and woman has died … on their side, of course! The cost might be high, but we will win. And when we do, we will have won their wondrous technology!"

- There are the business-exploiting points of view:

Wonk A says, "This ship isn't coming to make us part of some kind of galactic empire. If you look at it realistically, that concept makes no sense: The distances and journey times are too large and long. This means they want to trade. We should be preparing by researching they want."

Wonk B says, "Working out something exclusive with the ship doesn't seem possible at this time. But working out something exclusive with our governments is something that seems quite possible."

Wonk C says, "These aliens aren't going to be interested in gold or silver. They can get that kind of simple stuff anywhere. They are going to be interested in complex things … things that can't be just picked up off the ground anywhere. Living things are a prime example of what they are going to be interested in. Living things are unique and complex." C thinks a bit then says, "They may want women … er … I mean colonists … both women and men!"

- And there is the NGO point of view:

This is a NGO fundraising advertisement. It starts with a quick scene from *Star Trek* where the avenging alien whales come to earth. Then cuts to a famous NGO personality holding a fluffy kitten. She says, "The aliens are coming. Now it is more urgent than ever that we save the children and the animals. Please donate to St. Judas Hospital so that we can provide more mobility for our crippled children. They will need it while they are being taken care of in shelters that will keep them hidden from the aliens. And with the children, we will also send fluffy kittens to keep them company. They will save each other. Your donation will save both a child's and an animal's life."

- And where there are NGOs, there are scammers:

This is a robocall: "Hi there. I'm Prince Nabanooboo from Nigeria. I'm calling to appeal to you to help save our Nigerian schoolgirls from these terrible aliens. To make a donation, after the beep, please provide your name, address, bank account number, and how much you would like to donate."

- There are deniers:

Denier A says, "Pfft! This whole star ship business is a hoax. Don't believe it for one minute! The aliens are already here! They are holed up in Area 51."

- Of course, the politicians have their say:

Populist Politician A says, "First, let me say that all these ugly rumors about aliens coming for our women are just that: ugly rumors. We have no facts at this point about what they are coming for. However, if they are after our women, we will fight them to the last man! ... uh person! ... except of course for the children. We will keep them safer than the women. ... and the kittens, of course."

In sum, there is lots of cheap talk buzzing all around the solar system.

In addition to the cheap talk, there is also some reasoned talk. A panel of experts that includes Anthony, Clarissa, and John convenes on Myrtle's talk show. Behind them is an artist's depiction of the nearby star Gliese 581.

Myrtle starts, "Anthony and Clarissa, I understand we have a new discovery about this incoming starship. And you two are now newly minted xenoculuralists. This is a role we haven't needed before now."

Anthony says, "That's right, Myrtle. Now that we really have aliens, we need people who can study them. Clarissa and I have taken on that role.

"The astronomers watching the ship have now calculated that it is coming from Gliese 581. Gliese is a red dwarf star only twenty light years from the solar system, and it supports planets."

Myrtle says, "Interesting. What are the implications of that?"

Clarissa says, "These aliens are puddle jumping. Gliese 581 is one of the solar system's closest neighbors with rocky planets. They are moving from close star to close star. And it's only one ship. This is clearly not some invasion armada."

Anthony adds, "It's not an armada, but the ship could still be equipped with Godlike technology, and the aliens could be bringing Godlike terraforming equipment. They could be coming to make Earth into a Mars or Venus."

To this, John adds, "If they are, we are screwed no matter what we do. If they have that level of technology, they are way beyond our ability to influence."

Myrtle says, "What do you recommend?"

John says, "We should plan for trade, not conquest. If they have god powers and they want to exterminate us, they will, no matter what we plan. If they are here to trade, our planning can make a difference. A big difference."

<<<*>>>

For nine years, GO-1 changed as predicted by Foursmoots. It got brighter, and the spectrum reddened. Then, in 2073, something surprising happened. The first one to spot the change was Jimmy Golan, an intern astronomer in his twenties, working the night shift at Anthony's and Clarissa's observatory.

The night had been dull and quiet until he noticed the change on one of the screens. When he did, he immediately called Anthony. Anthony wakes up and groggily takes the call.

Anthony says, "Yes?"

Jimmy says, "Anthony. There's a change in GO-1. A big one. You may want to come over."

Anthony does and brings Clarissa with him. Jimmy is scanning the screens.

Anthony says, "What's up?"

Jimmy says, "The ship is about six months away and at forty percent of light speed, but look! It's gotten real weird!"

Anthony and Clarissa study the screen.

Clarissa says, "It's ten thousand times dimmer and oscillating in brightness. How can it be doing this?"

The trio keep looking at the screens. This is baffling.

Then Clarissa has an ah-hah moment "Wait! Isn't this something Foursmoots was speculating on? It could be corkscrewing its way in now instead of just pointing its engine steadily straight at us."

Anthony says, "If it is, start scanning for some kind of signal coming from it."

They do, and they pick one up. It is simple dot and dashes, like Morse code on an old-style radio. All three pair of eyes light up. This is big!

<<<*>>>

Foursmoots is on Myrtle's show again. He explains these latest developments. There is a screen behind him with an artist's animated conception of what is happening.

John says, "This change is happening because the ship was now coming in like a corkscrew rather than on a simple linear trajectory. The astronomers have also detected a communication signal coming from the ship."

Myrtle says, "A signal! What's in the signal?"

John says, "It is fairly strong, so the raw feed is being picked up all over the solar system. We have determined that there are actually four channels of information being sent. The first channel to be successfully analyzed is a basic 1+1 is 2, 1+2 is 3 mathematical sequence, followed by a galactic language equivalent of ABC.

"This is a Learn Galactic channel, and it cycles over about every twelve hours. People all over the world, including lots of schoolkids, are picking this up and deciphering it. Those who have figured out Galactic are using it to understand the second channel. This one is not nearly so easy to digest. We are all still working on it."

Myrtle says, "Have you figured out anything from it so far?

John smiles. He is proud of what has been accomplished. "We have. A lot. It describes more about technical Galactic language, and it is telling us how to build a new style of transmitter/receiver."

"Amazing!"

"This new style of receiver will likely allow us to understand the third and fourth channels. Those we don't have a clue about yet. They are indecipherable."

Myrtle expresses some concern. She is not alone in feeling this worry. "Are we going to build it?"

John shows mixed feelings, too, "This new receiver? You are not alone in feeling concern, Myrtle. Personally, I feel the sooner the better! This is most likely a trading channel. This shows that these aliens have come to trade. With the receiver functioning, we can finally get a lot of information about who we are dealing with."

"Could this be a tool to conquer us?"

"If it is, these aliens are way beyond our ability to resist. Here's how I look at it: We might as well look at the bright side because the dark side is too dark to imagine emerging from."

<<<*>>>

Six weeks later, there is construction on the moon of a new exotic-looking receiver and transmitter to talk on channels Three and Four. John explains what is happening, "Imagine building a 1960s-style FM radio from an instruction manual when your current state of radio technology transmission is doing an electric arc telegraph similar to what the Titanic had in 1912.

"You know some of the theory of what's going on with this advanced device, but according to the manual, you need roughly a thousand components that you don't have.

"Fortunately, the manual has the technical specifications for those components, and fortunately, they are all purely electronic components, not some mix of electricity and some mysterious Force X."

The transmitter/receiver is up and running by the time the ship is one month away from the solar system. Now the scientists—and the whole solar system—listen successfully to the two mystery channels.

While the world listens, John and Myrtle are talking again.

John says, "The two mystery channels are actually sixty-four channels each. One is sixty-four channels of Galactic Education Channel, and the other is sixty-four channels of Galactic Shopping Channel.

"The educational channel talks nonstop about Galactic history, social studies, science, and culture. The shopping channel talks nonstop about advertising things that the ship is offering to sell Earthlings and other solar system members.

"Both channels also have text channel equivalents that are downloading encyclopedia and sales catalog equivalents.

"We have sent back a greeting, and the two beginner channels are now being used for two-way communication. The aliens have, in effect, opened up two high bandwidth Internet connections, but connections with a one-month lag time ... just like we have hours of lag time when we communicate between planets. That, of course, is shortening as the ship gets closer."

Myrtle says, "Fascinating."

<<<*>>>

It is time for the military commanders to make some choices.

Colonel Sanders starts with a summary. "It has been roughly ten years since the alien ship was discovered and identified as such. Our solar system defense fleet has been created and now consists of ten carrier type ships, fifty manned fighter types, and over one hundred drone types of various sorts. In addition, the moons of Earth and Mars are now supporting defensive bases. We have had to prioritize, so we have not beefed up static defenses beyond Mars."

General Mayhem says, "So if these aliens stay out, say in Neptune orbit, what do we have to throw at them?"

Sanders says, "The defense fleet and the scout drones."

Mayhem says, "Not much, but then again, I guess that's a blessing."

General Destruction is a more hawkish sort. He sees no blessing. He thinks action is necessary. He says, "So, what sort of show of force are we going to present these aliens with? Are we just going to wave our hands and say, 'Hi there!'?

"How about we set off a nuke on Triton, just to show them that we do have some claws?"

Mayhem says, "And what if they respond by setting off some sort of ... I don't know ... Z-bomb on our Moon to show they have even bigger claws? I don't think so."

Destruction says, "Well, we should do something."

General Havoc says, "We should be sending the scouts."

Destruction says, "That's for sure."

Mayhem says, "Definitely."

Sanders says, "Are we in agreement on that? I will issue orders for the scouting to begin."

Destruction says, "Should we move the main fleet to Jupiter orbit?"

Mayhem says, "What's it going to protect there? Asteroids?"

Destruction, "It will be between us and them."

Havoc says, "Only half the time. The other half, it will be far, far away …"

Destruction interrupts, "Only if this goes on for years."

Havok finishes, "Or it will be burning tons of fuel so it doesn't orbit."

Mayhem says, "Let's keep the main fleet in the inner system for now. We can move it if the aliens move."

There are nods from all the generals. They have a plan.

The Big Picture

When the starship is first spotted, Earth is densely populated with both people and cyber. The rest of the solar system is sparsely populated. There are a handful of people on the Moon and Mars, and that is it for humans. Wandering throughout the rest of the solar system are about a thousand robot probes. They are all science probes. Science is the only reason to be off Earth. Other than communication satellites, nothing of commercial value has been discovered, so commerce is not swelling the solar system population or activity level.

As this incoming starship becomes more and more real, the bigger governmental organizations react by militarizing and calling for conferences on what to expect and how to coordinate Earth's response. The bigger supragovernment organizations also convene coordinating conferences. But getting these big conferences organized is slow and cumbersome compared to virtual and videoconferencing, so the big conferences become places to ritually approve the plans formulated and agreed upon earlier in the informal Internet-based conferences. It is good that these big conferences took a while to get organized, because agreement on courses of action didn't come easily.

The realists among the military argued that what was coming was some form of *War of the Worlds* space invasion. The pessimists among the realists argued that since the space invaders had a three-hundred-year technology lead, we had already lost and should concentrate on hiding and hunkering down. The optimists argued that 10 billion people and a whole Earth was a lot of people and resources. If there were a fight on Earth's surface, we would win. The dessert would be access to all that wonderful technology!

In the end, the big organizations that controlled the big resources decided to plan for both trade and invasion. And the smaller organizations all planned for whatever they felt like. The world was definitely in a panic, and there was a whole lot of end-of-the-world mentality being thought and spouted. What people came to blows over was how the world was going to end. Was it going to end heavenly or hellishly? Guessing the answer to this turned out to be something worth fighting for, mostly in backwater places that were not coping well with modernizing, anyway.

The good news was that the world had ten years to indulge in a lot of foolishness before the steadily brightening star did anything but steadily brighten. By the time the ship did something new, a lot of silly things had been done, but their silliness was being recognized as such, and the solar system community was starting to recover its levelheadedness. However, the process was far from complete.

The argument over whether to build the transmitter was intense, but it was decisively won by the let's-build-it side. The insurmountable justification: Build or not, the aliens would be here in six months. For the same reason, it was decided to build the transmitter/receiver on the Moon rather than some remote asteroid. If it were done on a remote asteroid, it would be finished well after the ship's arrival.

A month and a half after an Earth delegation sent back a "Hello" message, the two beginner channels changed to the standard Galactic formatting of channels Three and Four. Both were now available for two-way conversations. The aliens were, in effect, opening up two high-bandwidth Internet connections, but the connections had weeks-to-hours of lag time. Humans were familiar with this lag; there had been Internet to Mercury, Mars, and Titan for years now called Spacenet, so the alien channels were quickly integrated into Spacenet. The transmitter/receiver was the focus point: data from Spacenet were translated into Galactic there and vice versa.

By this time, it was also clear that the ship was going to end its initial approach at the orbit of Neptune, not somewhere in the inner solar system. The military had had ten years to prepare for this moment. The fleet was in top shape, the best shape it had ever been in. When it became clear that the aliens would stay coy, it was decided to take a watch-and-wait stance rather than launch a defensive armada to surround the incoming ship and escort it. The main fleet would stay at home, but in spite of yells and screams from the extreme pacifists, a few fast, lightly armed scout ships were sent to get closer to the aliens.

The alien ship replied to the Earth hello with its own on channel One of the reformatted beginner band and went on to explain that the aliens were delighted to encounter a new civilization here and that they wanted to trade. They continued on with an hours-long monologue asking questions about solar system resources and explaining their trading policies. They were clearly used to long-distance communication, so they expected that Earth people would be formulating replies and their own questions while they listened to this inbound message. They expected that the Earth people would begin broadcasting their reply as soon as they were ready, not waiting for the aliens to end their message. In Earth-Earth communication, this would be rude, but this is the way of long-distance communication.

It turned out "Galactic" was a misnomer for the communication protocol. According to the education channel, these people were Mirondians, and the solar system was on a distant edge of their cultural area, which was about three hundred light years across. There were other civilizations that the Mirondians were in contact with. The center of the Milky Way is about thirty thousand light years from the solar system, so Mirondian culture looked like Galactic only to Earth people who were making first contact. The Mirondian language being used between Earth and the spaceship was a scientifically designed, easy-to-learn pidgin Mirondian rather than a full language. Usage around the solar system was quickly updated from calling them Galactic to calling them Mirondian protocols.

Now that the solar system community and the spaceship could talk, it was time to see if fighting or trading would come next.

Chapter 2: Arrival

In The Trenches

The alien mothership settles into an orbit near Neptune. Thanks to Channel Three, it is now known that these aliens call themselves Mirondians. It is also known that they are not part of any galactic empire. The best they know, there is no such thing. The stars are just too far from each other to support such a governing structure. The people of this ship are their own bosses. They have no allegiance, even to their home world of Mirond.

Likewise, their language is Mirondian. They have been teaching pidgin Mirondian on channel Three. Once again, as far as they know, there is no Galactic language. The galaxy is a big place, and there are many, many different intelligent races and civilizations separated from each other by decades to centuries of space travel.

The solar system scout ships approach the mothership. They are greeted with a communication from it. Captain Ahab, a man in his forties, on the lead scout ship answers it, on the bridge along with a couple of crewmen. On the screen is Miro-John, a Mirondian. Due to the distance between the scout ship and the mothership, there is a four-second delay in responses. They are about as distant as the Moon is from Earth.

Miro-John's manner is pleasant. He does not seem arrogant or fearful. He says, "Greetings, Humans. Welcome to being near to our mothership. Please do your scanning and then move to a respectable distance—in your terms, one astronomical unit. We will interact further with you at the Show Room. We will set that up in the Jupiter orbit. We would prefer you to keep your distance from the mother ship."

(One AU is the distance from the Earth to the Sun.)

Ahab answers, "All right. We will make one close pass and then pull away. Where is this show room going to be?"

"The exact location is now being researched. We expect that to finish soon, and we expect to launch a ship for there in one week. I think you will find scouting that location will be much more interesting than scouting this one."

The scout ship fleet makes a pass by the mothership. After it does so and is flying away, the mothership becomes nearly transparent and loses color contrast. It doesn't disappear entirely, but it is hard to see. This is advanced electronic counter measures (ECM) in action.

Ahab says, "Hmmm, they do have powerful ECM. I guess this is what he meant by things getting dull out here. We will leave a drone that will stand off. The rest of us will head for Jupiter orbit."

The scout ships rocket off leaving the mothership. They are headed for Jupiter orbit. One week later, the drone reports that a craft has left the mothership and is headed for the inner system. The craft is monitored, and it heads for the Trojan asteroid cluster that orbits behind Jupiter.

<<<*>>>

Six weeks later, the craft reaches its destination and transforms. It picks one of the asteroids there, and using that for raw materials becomes the Show Room asteroid for the Mirondians.

This generates more talk show fodder for humans, and Myrtle and John are right on top of it.

John starts with, "Oh … and it turns out 'Galactic' is a misnomer for the communication protocol. According to the education channel, these people are Mirondians, and the solar system is on a distant edge of their cultural area, which is about three hundred light years across. There are other civilizations that the Mirondians are in contact with.

"And, the Mirondian language that is being used is a scientifically designed, easy-to-learn pidgin Mirondian rather than a full language."

Myrtle says, "What happens next?"

"There is a lot of trade talk going on even as we speak.

"Even more exciting the ali … er, Mirondians are setting up a show room in one of the Trojan asteroid clusters in Jupiter's orbit. The ship arrived last week. We expect the show room to be completed in six months to a year."

John blanks out a moment, listening to his communicator. "Wait! This is astounding news. This show room is nearly complete. The scouting fleet is just arriving in the area, and we have videos."

On the big screen behind him is a scene being taken by a scout ship. It shows that the show room ship is next to an asteroid. There is a lot of hustle and bustle going on between the ship and the asteroid. There is a building complex being assembled on the asteroid. This will be the Mirondian show room.

The construction is happening so quickly that it looks like time-lapse photography.

John is watching this with dropped jaw. He is amazed.

Myrtle says, "What are you seeing, John?"

"It's … It's … It's happening so quickly! It's like 'blip,' and there is a building there. This is nothing we expected."

"What does it mean?"

John smiles. "It means there are still lots of surprises coming."

<<<*>>>

The building complex on the Show Room asteroid is complete, and human ships have come to drop off human dignitaries. Three of those are Anna Hatchet, Paula Stove, and Mickey Palusi, all in their forties and all of whom are veteran trade negotiators.

Inside, they meet with Miro-Sam, Miro-Alex, and Miro-Bob, also seasoned negotiators, but not with humans. This human-Mirondian negotiating is brand new for both sides. The humans are wearing fashionable crew suits. These are considered high fashion in solar system business circles these days. Purely Earth-bound business circles have a different fashion. The Mirondians are wearing high-fashion designer crew suits, designed for mother ship living. Their crew suits put the human suits to shame because of the fabrics and materials used, but they are not ostentatious. This is just more of the Mirondians showing off their stuff, as in, what they have for trade.

The humans are led through an atrium that takes their breath away. This is showing off all sorts of stuff to impress the human visitors and fire their imaginations as to what they want to trade for. It is much like a pavilion at a world expo. The Mirondians take their time leading the humans through this exhibit hall. They let the humans drink it up.

Miro-Sam says, "These are some of the things we have fabricated on the ship that may be of interest to you. Feel free to look them over."

Anna says, "You made these on the ship? You didn't bring them from another star system?"

Miro-Sam says, "We have manufactured them. We want to give you an idea of what we can provide in large quantities."

Paula says, "So your ship is a factory as well as a mobile trading post?"

Miro-Alex says, "Indeed. In fact, it is more factory than trading post. We do a lot of value-adding, I believe you call it, as we travel from star system to star system. We have a lot of time on our hands."

They move from the atrium into a meeting room. It is as fantastic as the atrium, but in a subdued way. This is a place designed for serious trading talk. The Mirondians offer the humans refreshments and, once again, what they offer is neat stuff.

Mickey says, "This is fantastic. You do food well as well as the other things we have seen. How did you learn to make this food?"

Miro-Bob says, "We have been watching your entertainment videos. All metabolizing creatures pay lots of attention to their food. So as we approach every system, we carefully observe those processes."

Miro-Sam adds, "But your world here … Earth … is quite exceptional."

Anna says, "Our home world?"

Miro-Sam says, "Yes. It is quite rare to find a planet so lush with life. Most are more like your Mars or Jovian moons. They have pockets of life in small, favorable niches. This business of living things spreading all over the surface is rare … and quite fascinating to us."

Miro-Alex says, "A world with 9 billion people that is thickly covered with life from pole to pole. You have no idea how rare that is in our part of the galaxy!"

The small talk continues for just a little while longer before Miro-Sam says, "Shall we get down to business?"

They all sit around the table.

Miro-Sam continues, "You have been through our catalog, right?"
The humans nod.

"So let us talk about possibilities that are beyond the catalog. There are two things that come to mind right now, but there will be more in the future.

"The first is that we would like to get you moving around this solar system faster. Your infrastructure is primitive. Therefore, we would like to talk with you about constructing factories to make Mirondian-style propulsion systems."

The humans look at each other. This is huge.

"Second, we would like to help you identify and exploit your small moon and asteroid resources better. The fact that this show room was built at a Trojan asteroid point that was previously uninhabited is a striking example of under exploitation."

Anna says, "These are valuable offerings, indeed. What do you want in return?"

Miro-Sam says, "We want a share in what is produced, and we want to visit Earth. We want to experience your … jungle planet … which it is from our point of view."

Mickey says, "These projects are going to transform our world as we know it and produce billions in profits. I don't think doing a little tourist hosting on the side is going to be any problem."

The Big Picture

In addition to face-to-face negotiating, once greetings were exchanged over the communications channels, there was a lot of "meat" in the form of information exchange and trade opportunities. These channels were open to all, so the entire solar system community was caught up in discussing what came next. A lot of what got talked about was specific trade deals. Basically, each time Earthlings or Mirondians expressed interest in a particular possible project, it was assigned to a communication channel. Channels One and Two were kept open for general discussions. The most discussed project early on was where the Mirondians should set up their show room. From the solar system point of view, there were several conflicting desires:

- Somewhere that would let the solar system military defend it well if trade turned to blows
- Somewhere that was easily accessible to Earth, still the population and cultural center of the solar system community
- Somewhere that was easily accessible to the rest of the solar system community

Defense considerations and access by the solar system community ruled out Earth and the Moon. Intense radiation and a deep gravity well ruled out the moons of Jupiter. It was finally decided that an otherwise nondescript asteroid in the asteroid belt between Mars and Jupiter was the best choice from the solar system perspective. This choice was passed on to the Mirondians. They first asked if something in the Neptune orbit range might be better, but after the next exchange, they agreed with the solar system choice. "We hadn't given good consideration to the current state of your propulsion technology," they explained. But a small change was made: An asteroid in Jupiter's Trojan asteroids was used instead of one in the Asteroid Belt.

The official Earth delegation launched and arrived six months later. While they were headed there, a dozen other ships from all over the solar system community, including more from Earth, launched and made their way to the show room. The show room was going to be a busy place, indeed! All the while, the ships were approaching, and the communications lines hummed.

More Mirondian-compatible transceivers were built. Once the concept was understood, they were cheap and easy to build, and it wasn't long before they were valuable for all sorts of solar system communication needs. Even if the Mirondians were to vanish before a single trade had been transacted, they had started a communications revolution in the solar system. Given how easy it was to make transceivers, it was only weeks before the government gave up on trying to regulate communication with the Mirondians. They were solar system superstars. Millions of people wanted to talk to them, and there was plenty of bandwidth to do so. The government decided to let the Mirondians handle their own fan-club problems. The government would monitor, not regulate.

Solar system researchers went back to the Mirondian encyclopedia they had downloaded over the text channel and did their own research. They reported, "Most settled planets Mirondians know about are Mars equivalents, planets or moons with mildly life-hostile environments that have been lightly terraformed so that their inhabitants can live underground or in domed cities. Planetary populations tend to be in the 1-to-10 million range. By those standards, Earth is a rare, primordial jungle planet. No wonder they are so curious."

The value of the trade conducted was simply enormous. It seemed as though anyone who got any sort of trading deal completed with the Mirondians came out a billionaire, and it seemed like everything that was traded for was of potato magnitude in its potential effect on solar system culture. (The potato was one of the wonders that came to the world from South America after Columbus discovered it. Who hasn't eaten French fries? Who can imagine a world without them?)

Chapter 3: Warts

In The Trenches

It is now 2076, two years after the Mirondian mothership went into orbit near Neptune. There have been lots of changes happening all around the solar system. One of the surprising ones is happening at a tropical beach on Earth.

It is picturesque beach next to a nice resort. There are gentle waves and sand on one side and lush jungle vegetation on the other. It looks pristine in one direction, but in the other direction is a high-class resort with cabanas and so on. There are people enjoying the beach. It is a very usual looking scene.

Then it gets unusual. Our Mirondians from the meeting room, Miro-Sam, Miro-Alex, and Miro-Bob, come out of the hotel to enjoy the beach. However, they are dressed in environment suits. The weather at this tropic beach is fine for humans, but not Mirondians. In their suits, they frolic and beachcomb much as humans do. They admire the beach, the sky, and the water, but what they admire most is the jungle. That dense, lush, living greenery fascinates them.

Some of the humans stare at them when the Mirondians have their backs turned. Some stare in amazement, but some stare in anger that they are there.

<<<*>>>

Beach time is over. It is back to trading. Anna, Paula, and Mickey are back with Miro-Sam, Miro-Alex, and Miro-Bob in the Show Room asteroid meeting room.

Things have changed over the last two years. Anna, Paula, and Mickey are sporting new suits, and these new suits have a lot of Mirondian technology incorporated into them. This is just one example of Mirondian inspiration spreading through human culture.

The Mirondians have changed too. They are sporting crazy-looking tans and Hawaiian shirts that they picked up while visiting the beach. These are fashion changes inspired by their visiting Earth as tourists.

They start with small talk.

Miro-Sam says, "Those beaches you have … the ones next to oceans … in the hot areas? Those are just incredible! I can see why you humans line up to get to them. And you … swim in that water too? Amazing."

Paula says, "I'm happy to hear you have been enjoying the sights. Yes, we humans love those beaches. And we travel to experience many other environments as well."

Mickey says, "And thanks to the improved propulsion systems you have shown us how to build, we are now developing tourist industries all over the solar system. For us, that's amazing."

Anna brings the small talk to a close. "Shall we get down to business?"

Everyone sits at the table. Graphs spring up on the displays around them. They show dozens of business progress reports.

Anna says, "As you can see from the progress reports around us, the progress has been astounding, but there are two items we need to talk about.

"First, there has been a lot of social unrest among some human societies. The progress has been … uneven. Second, we want more. We have been, and remain, most impressed with how quickly you Mirondians can get things built. We want you to teach us. We want not only the blueprints for these astounding inventions, but we also want to learn how you do things. We want to master these blip construction techniques."

Miro-Alex says, "'Blip?' This is not coming up in my dictionary."

Anna says, "It is a new slang word. It means Mirondian construction techniques. It originated because humans have watched you make things. You start and … blip … it is finished! … Blip construction."

Miro-Alex says, "Ah … I like it! You humans are surely inventive with your languages."

Anna says, "So … what would you like from us in return?"

Miro-Bob looks at the others and then confides, "This relationship has gone well from our point of view too, Ms. Hatchet, and we have started to dream big as you have. There are a lot of resources in this solar system, and there are a lot of you humans inhabiting it. This means we can dream big. And for space traders such as ourselves, this means dreaming of establishing a mother world, a place that can spawn dozens to hundreds of starships. Each can then go off in its own direction to explore its own venue of distant stars. This solar system, with all its resources and humans, has the potential to become that."

The humans look at each other.

Anna says, "Exploring the stars has long been an aspiration of many humans."

Miro-Bob says, "It has? I find that surprising."

Paula says, "Ever since we have been human, we have looked up at the stars at night and wondered what was out there."

Miro-Alex says, "Ah … that's right! You do have those clear night skies. So unusual for a jungle planet."

Miro-Sam says, "It seems we have some more aspirations in common. We may be able to move this Mirondian-human relationship yet another surprising step."

Anna says, "We may, but before that can happen, we need to solve the problems of deep human unrest we are having today."

Miro-Sam says, "Ah yes. We have been noticing that as well. In truth, we have found the extent and deepness of it surprising. It doesn't match our projections well. But then again, our projections did not take into account your enthusiasm for interstellar exploring. How can we help?"

Paula says, "We're not sure. For us, this unrest is expected. It happens each time there is deep economic and technical change on earth, and your coming and trading with us has certainly provided that!"

Miro-Alex says, "What will come of it?"

Paula says, "New ways of human thinking, and new ways for humans to do things. The problem is that we humans don't know ahead of time which ways will be selected. Exploring and selecting is what the unrest is all about. But this exploring and selecting scares many people. This is why we have the unrest."

Miro-Alex thinks a bit and then says, "It is … a time of chaos then?"

Paula says, "Very much so."

Miro-Alex says, "Then I think we can do a lot. From our own social studies, we know that chaos is a time when small pushes in the right places can make big differences in final outcomes. Between us, let us decide what outcomes we want for this chaos and investigate what tools we have for pushing the results in that direction." He looks around at the other Mirondians. "Let us confer among ourselves, and let us confer more with you."

The Big Picture

It took only two years for some of the warts to show up from this trading bonanza. The root of the problem was the divide between those who could quickly adapt to these new opportunities and those who got scared or uprooted by the disruptive changes.

There were also the ongoing mysteries of how the Mirondians did what they did. For example, one was the mystery of the size of the Mirondinan crew. This mystery came to light as the human-Mirondian projects grew in size and extent.

The Mirondians were happy to outsource. There were things they wanted built that took a lot of labor and resources. A lot of what they outsourced was sold to the solar system community. They provided expertise, and the solar system provided labor and raw materials.

The magnitude of the outsourcing steadily grew over the first three years that the Mirondians were in the solar system. The number of projects grew and grew, and at first, this steady growth was a mystery to the Earth negotiators. They would conclude a large joint construction or engineering venture, and the Mirondians always seemed to have manpower available—alien power, actually. The deal would be concluded, and shortly thereafter, a ship would leave from the mother ship with ten, a hundred, once five hundred, Mirondian Surface Engineers (as the Mirondians called them, MSEs as solar system people called them), ready to add their expertise to the project.

Where were all these alien people coming from? The scout ship's pass by the Mirondian ship way back when it first arrived gave solar system people a good estimate on the ship's size. It was really big, the size of a small asteroid, but even that big, there would be a limit on crew size. After a couple of years of watching this steady stream of aliens come from the ship, it looked as if Mirondians were packed on board at roughly commercial airplane density. This made no sense for a ship that would spend years and years floating among star systems.

Another oddity was that these MSEs seemed to come from different cultures. The newer MSEs didn't know the same things that their bosses knew. The solar system people found this out as they worked with them on their various projects. A few were real sharp and knew what was happening from day one of a project. Most acted for all the world as though they were immigrants who had just sailed from some foreign land and were getting off the boat for the first time. They had no knowledge of what had been happening since the Mirondians had come to the solar system.

When the solar system people finally asked the Mirondians about this seemingly inexhaustible personnel supply, the Mirondians answered, "The MSEs? Why, they're colonists, of course. Civilized beings are one of the most versatile cargoes that we carry, and that makes them valuable. When we depart, some of your people will be invited to become MSEs. I assure you that it's a most rewarding experience."

"And just how do the MSEs travel?" asked the Earth representative.

"Oh, that's one of our secrets, something I can't tell you about at this time."

Yes, the Mirondians kept secrets.

Chapter 4:
The Big Announcement of Year Four

In The Trenches

One of the spectacular accomplishments of the Mirondian era was the building of the first space elevator. As it became recognized how expensive it would always be to use rockets to get mass out of Earth's gravitational well, physicists and space engineers started proposing alternative methods. In the 1960s, the space elevator concept gained a lot of popularity. In practice, though, the elevator and all the others proposed turned out to be even more difficult to accomplish than sending up rockets.

That was until the Mirondians added some of their technical wizardry. Then ... blip!

The project begins with another Mirondian ship leaving the Show Room asteroid. This one is filled with more MSEs and heads for Earth's orbit. Once there, the MSEs are close enough to Earth that they can use their blip-management techniques while remaining on the ship as they coordinate with human workers on Earth assigned to the space elevator project.

On Earth, a few of the newly hired engineers are gathered in a big meeting room to get introduced to the MSEs. They are of diverse backgrounds and getting to know each other as they are preparing to get their first instructions from the Mirondian who will be supervising them. He will be on a screen talking to them from the engineering ship orbiting Earth.

Three of the workers are Jason Furlong, a good-looking man in his thirties, Zeb Noches, a bit heavier and shorter than Jason, also in his thirties, and Judy Alonzo, a good-looking woman in an engineering way in her twenties. The three are chatting and drinking coffee when the main screen lights up. They sit next to each other at the large conference table with the others and wait to hear the first news.

Miro-Robert, the chief Mirondian on this project, appears on the screen. In this first appearance, he is new at this talking-to-humans business, but he improves rapidly over time.

Miro-Robert says, "Humans … I mean, ladies and gentlemen. You can call me … I mean, I am … Miro-Robert. … I apologize for my language. I am still learning your culture.

"I will be your supervisor on this space elevator project. As outlined in the job proposals you read before joining this project, your mission is to design the tools that will make the structural members of the elevator tower. These members have to be … smart … I think you call it. They have to change their supporting characteristics in response to several different criterion, such as load, and on the lower elements, weather conditions. Is this clear?"

The humans nod their heads. It takes a moment for Miro-Robert to figure out this means understanding and agreement.

"I will send the beginning information to your personal communicators. Use those to send back questions or information. Thank you."

With that, he signs off. Moving quickly from task to task is part of blip culture.

The speed of his exit surprises some of the engineers.

Jason says, "That was fast."

Zeb says, "Yeah!"

After Miro-Robert leaves the screen, information starts coming into the cyber links of the engineers. As they finish absorbing that, they get back into the real world and start chatting with each other again.

Judy says, "Lunch?"

Jason says, "Sounds good to me."

The trio head out for the food court. They eat and chat some more.

Judy says, "This is going to be quite a project."

Jason says, "That Miro-Robert guy is certainly fast-presenting."

Zeb says, "I think that's part of being blip. I'm real curious about that."

Judy says, "Me too."

Jason says, "Well … when it shows me how to get laid faster, then I get interested. Sorry, Judy."

Judy is not bothered.

"Zeb and I, by the way, go way back. We went to high school together."

Judy says, "Really? Where was that?"

Jason says, "Lexington, Kentucky."

Judy says, "I'm a Colorado girl, myself. Boulder."

<<<*>>>

The trio work hard together, and well, but the task is a challenging one. It is challenging for many reasons, but one is that Miro-Robert pops in frequently on the communicator to change details of their mission. If he were a mortal Earth manager, this would be frustrating and demoralizing. Moving goal posts are never fun. But because this is being done blip style, the changes quickly make sense and help, and the project keeps moving ahead at a blistering pace.

In just a month, they have their part of the project finished. They are in their office admiring a component that has just come off a 3D printer.

Jason says, "Wow! We started this when? Just four weeks ago? This should have taken a year and ten people."

Zeb says, "This blip stuff is impressive."

Jason says, "And so are we!"

There are fist bumps all around.

Miro-Robert appears on the screen. He says, "Good work … excellent work. I'm sending you the next phase." With that, he is gone.

<<<*>>>

It is eighteen months later, and the space elevator is now complete. Cargo units are moving up and down. A huge breakthrough in Earth's ability to reach space is now in place and functioning.

There is a celebration party in progress. Miro-Robert comes down and joins in. He wears an environment suit. He moves quickly from group to group. It is as if he is engaging in blip socializing. The trio, on the other hand, are relaxing and winding down.

Judy says, "So ... what comes next for you gents?"

Jason says, "There's a new factory starting up, and I have a job there. It will make components for this new intrasolar system propulsion system the Mirondians have sold us."

Judy says, "Neat! What about you, Zeb?"

Zeb says, "Oh ... I'm real curious about how this new build-it-so-fast system works. Real curious. Miro-Robert is going to be teaching a class on how to do blip. I'm going to be taking that for a while."

Judy says, "Neat."

Zeb says, "Neat, but real intensive. It will keep me occupied full-time."

Judy looks at Jason, "Not for you, eh?"

Jason says, "Nope, not for me. I figure if I learn blip, the first thing I'll use it for is drinking."

They all laugh at that.

Jason says, "And how about you, Judy?"

Judy says, "Me? I'll be staying with the company. Now that we have it figured out, this smart girder business is finding lots of other applications. I figure I'll be on Earth doing this a couple years. Then ... who knows? Maybe on some colony up above. I've always wanted to try living in space."

Zeb says, "Wow! That's high ambition!"

Once again, there are laughs around.

<<<*>>>

Two years later, there is a conference on the Moon. Zeb and Judy run into each other there.

The climate control is suitable for human beings—plenty of air and the right temperature—but it is only at Moon gravity, so people bounce when they walk and handling objects such as plates full of food is different.

Zeb says, "How are things going for you, Judy?"

Judy says, "Oh, just fine. Really well, in fact. I've just been promoted. I'm now Base Chief at the Belt 226 Transfer Station."

"Really? Congrats! If I'm not mistaken, that's the one that is handling the new cryonics nanoparticles coming from Io, isn't it? Quite a responsibility."

"You got it. It's tricky stuff to deal with, but we're getting the hang of it now. And how about you?"

"Oh, I'm on my third start-up right now, still on Earth. The first one fizzled out pretty completely, but the second one did well, and selling out of it let me get this third one started.

"This one is looking hot, white hot. We will be making a new drug, something called Miracoxin. It will be a super cold pill. It works directly on the human immune system, not the virus or other foreign agent that is firing it up. Take it, and flulike symptoms will vanish."

"What happens to the virus?"

"Oh, the immune system is still functioning, but it's not going overboard. That's the difference. The virus will take a few days longer to be contained, but the person will be feeling normal while it happens. Plus, there will be other drugs that can work on the virus. The real key is that Miracoxin makes people feel better as soon as they start taking it. That's why it's hot."

Judy smiles at him. "It sure sounds like it. Congrats in advance."

Just then, they look at one of the display screens on the wall showing a newscast. They notice that it is news about some kind of protest happening on Earth, and there is something happening there that is special for them.

They say together, "It's Jason!"

The protester being interviewed is Jason. They now pay more attention. They watch him and give him a call when his interview ends. This shows up on a screen that both can see and talk to.

Zeb says, "Hey Jason! What's up, ol' man? We just saw you on the news."

Jason says, "Oh that! These days, I'm the chairman of the Cleveland Mirondians Go Home committee. I lost my job at the factory because they are relocating engine construction to space stations. This is an outrage."

Judy looks at Zeb with a bit of surprise. She whispers to him, "Ouch! That's one of the projects my group is picking up." Then she says to Jason, "Why is that an outrage? Those engines get used in space. Why not make them there?"

Jason doesn't have a quick answer, but he comes up with, "Well ... umm ... because it cost me my job! That's why."

Judy says, "So ... what are you going to do next?"

Jason says, "I'm not sure. I'm not sure what's going to open up here on Earth. The good news is that I'm entitled. I'm not going to starve or go homeless. But ... man! I still want a job. And I think it's those Mirondians who are cutting me out of one! That's why I'm protesting."

Judy says, "You seem to be doing a good job at that. We spotted you way up here on the Moon."

Jason looks a little bug-eyed at that. "That's where you are? Both of you? Wow! Things have sure changed."

Zeb says, "I'm working on Earth. I'm just up for this conference."

Judy says, "Well, keep looking. I'm sure something will turn up. These are quickly changing times."

Jason says, "Given how protester work is about selling something. I'm looking into that now. Bye now."

They break off contact.

After they do, a Mirondian comes up to them. It is Miro-Sam, who looks and sounds quite different from Miro-Robert.

Miro-Sam says, "Judy, before you leave, please come see me. It's routine, about something at the base, and a quick talk here will get a solution started."

He is about to move on when Judy grabs his arm and says, "Wait. I'd like you to meet a friend. This is Zeb. We worked on the elevator together."

Miro-Sam says, "Nice to meet you, Zeb. Perhaps we will get a chance to work together in the future."

He bows and moves on. He is Mirondian: They have met, and he gets on with his business.

Zeb says to Judy, "Mirondian ... but way different from Miro-Robert."

Judy says, "You noticed it too. It feels like they came from different cultures ... maybe even different planets. How could that be?"

Zeb shrugs, "There's still lots of mystery out there, I guess."

Miro-Sam is walking by again. Zeb stops him, "A question for you, Miro-Sam, if you have the time."

Miro-Sam says, "Please, ask away."

Zeb says, "Are you from the same place as Miro-Robert? You two seem so different in how you communicate."

Miro-Sam says, "We are not. We are both colonists. We come from different worlds."

Zeb says, "Different worlds! How did you both get on the same ship?"

Miro-Sam says, "Like I said, we are both colonists. That is all I can say on this right now. Good-bye again." He continues on his way.

Zeb says to Judy, "Yeah, I guess they still keep some secrets."

They go on about their conferencing.

<<<*>>>

One year later, there are more surprises.

This time the center of surprise for Jason, Zeb, and Judy is in the halls of US Congress. Jason and Zeb are called to testify before a congressional hearing.

They are in a hearing room with all the usual hustle and bustle that accompanies a hearing—committee members, aids, witnesses, and an audience.

The hearing chair, a man in his fifties, starts off. "We will now continue our investigation of Miracle. This is a new recreational drug that may be yet another threat to our well-being that has come from the Mirondian ship. Our next witness is Zeb Noches, president of Miracoxin Inc., maker of Miracoxin."

Zeb speaks into his microphone. "Our company acquired the rights to make Miracoxin from the Mirondians. Miracoxin suppresses the cold symptoms that a person suffers when they are having flulike symptoms from any source—flu, cold, drug reaction, whatever. It's a cold tablet on steroids and as safe as aspirin. Our company has been making it and selling it, and it has become popular the world over. However, six months ago, some of those bad apples in the druggie community discovered that if you acidify Miracoxin before you take it, it develops euphoric properties as well. It makes you feel good as well as relieving cold symptoms. This modified form is called Miracle. I would like to point out that this is not a use that Miracoxin Inc. discovered or recommends. It is also not something that affects the efficacy of Miracoxin. This is not something that should affect our ability to help the world alleviate itself from cold symptoms. Thank you."

The chairman says, "Thank you, Mr. Noches. Our next witness will present a different point of view. Mr. Jason Furlong, will you give us your point of view?"

Jason talks into his microphone. "Miracle has opened a new front in the war on drugs. It's not as though we didn't have enough problems in this arena already. We should not blame Miracoxin Inc. for this problem. Rather, the ones to blame are the Mirondians! If the Mirondians hadn't shown up, we wouldn't have this … Miracle … to worry about. Thank you."

The chairman says, "Thank you, Mr. Furlong. We will now recess for ten minutes."

The hearing breaks up. Zeb and Jason leave, but they get together again in a restaurant at the transport terminal where they are both catching rides home. They catch up with each other.

(Note that with driverless cars, buses, and airplanes, transportation is done a lot differently from how it is in the 2010s. This terminal handles taxis, buses, and airplanes, but light rail systems are now an anachronism for passenger use. They are too inflexible in routing to be of much value.)

Jason says, "It looks as though it's been a busy year for you, Zeb. That start-up seems to have hit big time for you."

Zeb says, "It has. Keep it under your hat, but we're planning on going public, and if it goes as planned, I'll be a billionaire."

"For real? This Mirondian business has treated you right, that's for sure."

"It has. And apparently, this year has been a busy one for you too, Jason. From man on the street to congressional testifier. That's quite a move up … in an odd path … I'd say anyway. How'd you wrangle that?"

"Keep it under your hat, but Miracle has done well for me. I hooked up early on with some of those bad apples, as you called them. I've got me a distribution network!" He runs his hands up and down under his suspenders. "I've bought myself some legislators. I'm no billionaire, but I'm making a living. But I wish things were going differently. I really do wish those damn Mirondians hadn't showed up. I wish they'd take a hike. I'd really like to be doing some real work, not the corner whoring I'm doing now."

"I hear you, but I don't agree. Man! We are making such progress these days."

Jason is about to say more, but their respective transports arrive, and they part ways.

<<<*>>>

Judy is attending a class. Her professor, a casually dressed woman in her forties, is explaining why the human community has become restive since the Mirondians arrived.

"Tension in the human community is high and growing because the things that Mirondians trade to the solar system community are not as trivial as chests of gold or bags of diamonds. They are much more earthshaking. Think of it this way: If a man comes to a town with a chest of gold and starts spending it like a young sailor on leave, he is happy and the town's merchants are happy, but there is little change to the daily life of the town. The gold is nice for the town's well-being—it makes the town better—but it doesn't change the day-to-day life of the town. The town lives on and simply waits for the next man with gold to come along.

"But if a man comes to town, and it has only people and horses for transport, and he builds a big bicycle factory that sells cheap bicycles, then he deeply alters the town's day-to-day living patterns.

"He provides new kinds of jobs, a new style of transportation, city growth, city pollution … lots of changes. Some good, some bad, and many that are both.

"In short, he changes how the town lives. Some townspeople are going to be happy about the change, and some will be upset.

"It is the bicycle factory kind of change that Mirondians are bringing to the solar system. The changes we are experiencing are huge."

Judy raises her hand, and the professor recognizes her. "Speaking of transportation revolutions, that's what I'm involved in. My company is making some of those new constant acceleration-propulsion systems."

The professor says, "The ones that are cutting Earth-Mars travel times from years to weeks?"

"Yes, plus making the trip healthier because the gravity stays higher while the trip is in progress."

The professor asks the class, "What other changes will this bring?"

Another student answers, "That, plus the space elevator, mean a whole lot more people are going to be leaving Earth."

"Leaving in what sense? Vacations?"

"No, as in emigrating—going to other places to live and work."

"What's another change that is needed for that to happen?"

"We need to be able to quickly build infrastructure ... blip fast. That's what I'm involved in. We are projecting growth from the thousand off-Earthers when the Mirondians arrived to more than a million by the end of this year."

"That many?"

The student nods.

"Yes, indeed. Our world is changing quickly. And I predict even more surprises are coming."

<<<*>>>

Not only humans are prospering. The Mirondians are prospering as well. This means more vacation time for them, and they are spending that time where they love it—at the beach.

Miro-Sam, Miro-Alex, and Miro-Bob are back at the beach in their environment suits. The suits are improved. The trio can now splash around in the water as well as wander the beach.

They approach the jungle, curious about it. As they do, Irving Eggelston, an eccentric British birdwatcher in his thirties, pops out with camera in hand. He is surprised as well.

He says, "I say! You're not from around here, are you?"

There is a pause while the Mirondians figure out what they have just heard.

Miro-Sam says, "We are from the Mirondian ship."

Irving says, "Ah ... that explains the suits."

Miro-Alex says, "We are here to look at the jungle."

Irving says, "I'm here to look at the birds in the jungle. They are so colorful. When I see them, I shoot them."

The Mirondians look at each other. This seems barbaric to them. Eggelston notices their concern. He holds up his camera. "With my camera, that is. I don't kill them. Would you like to join me as I look for more birds?"

The Mirondians agree to this, and the foursome walks into the jungle with Eggelston at the lead. The foursome all have a good time and the Mirondians get to personally experience the lush vegetation with lots of explanation about the various plants and animals they see. Eggelston even convinces them to sample some tropical berries. They survive that, and they will have plenty of admirers when they get back to the ship.

<<<*>>>

It is New Year's Eve. It is Year Five since the Mirondians arrived. There is a gathering of thousands of people in a city square. But more than the new year brings them. The Mirondians say they have a big announcement to make.

Miro-Robert shows up on the big screen to make an announcement to the crowd. He is beaming. This is a joyous announcement for him.

"We are very pleased with how our trading and relationships with this solar system are progressing. We are delighted at how generous and responsive you people of the solar system have been. Thanks to your wholehearted cooperation, we have both prospered mightily. We are so encouraged that we have decided to start what for us is a most precious project: the building of a new star ship. This is not a project we take on lightly. And we do it only because we think Earth and the solar system can add so much value to this endeavor.

"Congratulations to you people of the solar system. You have proved to be wonderful trading partners."

The people watching are speechless, but they do not share Miro-Robert's joy. They are speechless with surprise.

Jason, Zeb, and Judy enter a restaurant and wait for a table. Judy wears an exoskeleton to help her deal with Earth gravity. She has become acclimated to the low gravity of the space colonies, and moving around on Earth is now a big effort. They chat as they wait.

Jason says, "What brings you down to Earth these days, Judy? It must be something special. You look like Iron Man."

Judy says, "It is. I'm here for my dad's funeral."

Jason says, "Sorry to hear about that. I guess it's a double ouch, actually. You have to wear that."

Judy says, "Yeah, I admit that I've gotten real used to being above the clouds. Being down here is a lot of work for me now. I do most of my Earth meeting through a video connection."

Zeb says, "Are you going to be okay? Oh, and sorry about your dad."

Judy says, "I'll manage, and I'll be real happy when the elevator is taking me up again."

They get a table, sit down, and order. Judy starts off the talk about the big news. "Well, boys, what do you think of this big news the Mirondians dropped on us?"

Zeb says, "Are they planning on crewing it with Earth people? If I were ten years younger, I'd be out in a heartbeat."

Jason says, "You'd go off to who-knows-where and pay the Mirondians a pretty penny to take you? I guess it's a good thing you're an old, rich man now."

Zeb says, "You wouldn't go? What an opportunity."

Jason says, "If they start sending stuff away from here, stuff will come back too. Real strange stuff. Maybe not such good strange stuff. Mark my words! And in the meantime, we've still got lots of poor people here. We should be doing more to help the poor, not build some fancy-ass, interstellar spaceship."

Judy says, "This is big, and I'm mostly looking forward to the contracts."

Zeb says, "Think of the opportunity … Humanity can expand across the galaxy."

Jason says, "Pfft! I've heard that before … all the time in science-fiction movies."

Judy says, "Across the galaxy is a stretch, but at least a lot of places nearby. Even the Mirondians don't go across the whole galaxy. It's too big. And as best they know, there's no Galactic Empire, either."

Jason sums up his point of view, "Fix home first." He then changes the topic. "Say, you been following the Ram Jets lately? They just won their fifth in a row."

As they eat, they listen to open-mike night for comedians. Sally Forth, an aspiring comic in her twenties, gives her monologue.

Sally says, "In Mirondian eyes, this announcement is supposed to have been one of great joy for both Mirondians and solar system people. For them, it is like when a young bride gets dewy-eyed, cuddles up to her new husband, and whispers in his ear, 'Darling, I have something special to tell you.' He smiles back, hugs her, kisses her, and knows that their baby-making exercises have been successful. In solar system eyes, this is like when a dad's sixteen year-old daughter cuddles up to him and tells him she has had the same success with her high school boyfriend. The dad says, 'Whoa! Let's talk about this!'"

<<<*>>>

Six months later, Miro-Robert is on the Mirondian News Channel giving further details about the new Mirondian ship that is to be built. Diagrams come up to emphasize the points he is making. The diagrams show that the new star ship will be huge. This was going to take a lot of resources, even for a solar system which has an Earth-style planet in it.

Miro-Robert says, "Yes, it will be a grand ship. It will be five times bigger than our current ship. It is the size of a comet, and it will hold a medium-sized city's worth of people in it. It will be finished in five years."

The human commentators are buzzing about this announcement Miro-Robert has made. Myrtle is right on top of it on her show. Judy is on one screen as an off-world expert, Miro-Robert is on another.

Judy says, "There are far too few MSEs to get this project completed as scheduled. This means that Mirondians were going to have to teach even more blip construction- and management-techniques to a lot of solar system people. I don't know about how other people feel, but I want to be first in line."

Myrtle says, "Who is going to crew this new ship, and who is going to pay for it?"

Miro-Robert says, "We will issue contracts, as we always have. You, people of the solar system, will help us man the ship. It will be an honor for many of you. It will be your chance to see the wonders of other star systems and spread your seed to other worlds."

The Big Picture

Yes, the Mirondians kept secrets, but for the first five years, the distaste caused by the secret keeping was overshadowed by the spectacular benefits of the trading. But tension was growing in the solar system community, and as the fifth anniversary of the Mirondian arrival approached, it was high.

One example is the transportation revolution. Before the Mirondians came, mankind was moving around the solar system using boost-and-coast rocket technology. This was the same technology that got man to the Moon in the 1960s. Mankind was moving around, but the process of moving people and goods from one planet to another, or from a planet to an asteroid or moon, took weeks or years, and it was expensive. As a result, the off-Earth colonies were small, supported only by science aspirations and not commercial activity, and growing slowly. They imported only the most important things they needed and no luxuries. The solar system was Earth-centric. The Moon had the biggest off-Earth colonies because it was quick to get to from Earth, and Mars had the next biggest for the same reason.

The Mirondians traded to Earth a constant acceleration propulsion system that could sustain 0.5G acceleration to anywhere in the solar system. Using the blip construction techniques, a factory making these systems at the rate of one a month was available six months after the space elevator opened, and it was expanded to two a month at the end of the first year when it was clear that demand for these systems was huge. A year after that, the human-propulsion people made a breakthrough, and 1G propulsion systems became practical for luxury ships.

On these new ships, a trip from Earth to Mars took a few days to three weeks, depending on their orbital relation. Even more important, ships could launch at any time—1G acceleration was so powerful that orbital niceties such as optimal intercept times could be ignored. And with constant 1G acceleration, the bad health effects of low-G travel were avoided. Traveling around the solar system became nearly as comfortable and convenient as traveling around Earth.

As a result of this new ease and cheapness, first thousands and then millions more Earthlings were willing to travel in space and take up residence in colonies. This uptick in willingness to travel would have caused a huge housing crisis on the colonies, but blip construction techniques were reducing housing construction times by five times. The colonies were able to sustain huge growth. Roughly a million Earthlings became emigrants in Year Five, which increased the off-Earth population of the solar system a hundredfold. They emigrated because they were curious and because there were hundreds of new off-Earth mining, manufacturing, and construction projects being initiated thanks to new Mirondian products and technology. A million moved, but a million did not stay. Most went for a while, came back to Earth, and then went out again. But many stayed and then brought their families. Mixed in with the million or so Earthling emigrants moving back and forth were a few hundred MSEs from the Mirondian mothership.

The physical facilities were able to sustain this population surge, but the social facilities were overwhelmed. Before the growth of Year Four, the colonies were neighborly places. The colonial governments were informal, and everyone in a colony knew everyone else. There was a lot of hazard outside the colony walls, so cooperation was high and crime was low. Mirondian technology cheapened everything in a colony. Housing, food, life support, entertainment, social values: Everything got cheaper. Many pre-Mirondian residents of the colonies became deeply upset at the cheapening social values; many did not because they loved the new people and because they could now import frivolous things, such as flowers. Those who were made billionaires by the changes tended to say the problems were deplorable but bearable.

This transportation revolution, and the colonial social revolution it brought about, is one example of the hundreds of changes that came with Mirondian trading, and it is an example of how each change raised the tension in the various solar system communities. The tensions were high, but the wondrousness was also high.

The solar system community was probably not alone in feeling the tension. The Mirondians were trading because they were getting wealthy as well. We don't know much about their community, but it is hard to imagine that their community did not undergo some wealth stress as well. What is known is that the Mirondian community seemed to fall in love with Earth. The Mirondians came as tourists to visit Earth and kept coming back. They loved seeing lush life. They loved coral reefs, temperate and tropical rainforests, cities, and zoos. Mountain ranges left them cold. Interesting rock formations caused by erosion held their interest briefly, but life—thick life—was a magnet to them. They couldn't stay away. The rest of the solar system communities got somewhat envious of this, but there wasn't much they could do about it. What made this even more ironic was that the Mirondians had to experience Earth from inside environment suits. Their native conditions were not Earthlike. When they visited somewhere such as Yosemite, they stood out. They were the ones in the spacesuits, surrounded by the flows of casually clothed humanity.

The announcement of building a new mothership started a firestorm of discussion about the Mirondian/solar system relationship. Many solar system communities started seriously reviewing the question of who was getting what from the relationship. Those who were unhappy, passed by, or simply looking for some anger to exploit started talking about all those things that had gone wrong since the Mirondians had arrived. Since there had been a lot of changes, there were plenty of sour stories to tell. (There were plenty of good stories, but the people who were part of the good stories were usually too busy doing better to be good storytellers.) Here are two examples.

The first is a fairly simple one. There were chronic headlines of "[Community Name] unsettled by crime wave. Are new settlers to blame?" The article would then go into a discussion of various high-profile crimes of the community, who was moving in, and what good they brought, but that there were a few bad apples in the mix. It would end with some local politician or NGO leader campaigning for better screening of immigrants. As people got angrier, these balanced stories evolved into sensational stories about lurid crimes and immigrant gangs running amok.

The second example is the Miracoxin/Miracle evolution described above. After the Miracle discovery, the antidrug groups were spreading stories that Miracle/Miracoxin was not as safe as was originally reported. The antidrug establishment had a new target. Memberships in antidrug crusades tripled. The problem was that Miracle was safe. No reasonably neutral study would confirm the accusations of the antidrug movement. The result of this mix—an attractive drug and deep worry about its attractiveness—was a new battlefield in the war on drugs. The battlefield was blamed on the Mirondians. "If the Mirondians hadn't shown up, we wouldn't have Miracle to worry about," argued those antidruggers who were frustrated that they couldn't win on this battlefield.

These are just a couple examples of the controversies that grew out of the huge social changes that Mirondian/solar system contact brought to the solar system. The solar system was quickly getting to be a better place for humans, but those benefits were hard to digest, and they caused many people a lot of worry and stress.

To those humans who wanted to voice discontent, the new starship announcement became a lightning rod. The Mirondians were tightlipped on the starship issue, which made controversy easier. For six months after the announcement, the Mirondians had little further to say on it other than that they were planning it and that humans needed to be patient. People were not patient; they bellyached a lot, and there were protests building.

Chapter 5: The Moon Crisis
(The Crisis of Year Five)

In The Trenches

It is a year later, five years since the Mirondians arrived. It is moon meeting time again for Zeb and Judy. They are gathered at a banquet.

Zeb says, "How's the year been going, Judy?"

Judy facepalms. "Oh my! It's been a rough one."

"No contracts?"

"Plenty of contracts, but lots of difficult work and not much profit. The Mirondians are getting hard-nosed. The scuttlebutt is that this starship project is going to be a big, expensive one—even for them. They are squeezing."

"Ouch!"

"Yeah, but we're adapting. We're doing more work for human contractors. They are still paying like they did last year, and there are more of them. And how about you, Zeb?"

"Well, the hot news is that I got accepted to the charter class at MirU."

"What? That new university they are opening up? The one that's teaching hard-core blip?"

"That one, indeed! And you know who else is in the class with me? John Foursmoots."

"The one who first discovered the Mirondian ship was coming our way?"

"One and the same. And he's been busy since then. He's been helping redesign the Mirondian propulsion systems, making them work at higher G force."

"Impressive! On many counts."

"I'm headed there from here. That's why I'm up here now. MirU is at the Show Room asteroid."

"What's Jason up to these days?"

"Funny you should ask: I was talking with John Foursmoots, and apparently he and Jason have crossed paths on Earth and are now buddies of some sorts."

Judy shakes her head at this news, "It's a strange, strange world we live in. What I heard was he has kept up his protesting and got tossed in jail a couple times."

"Foursmoots and jail time for Mirondians Go Home protesting? That is a strange mix."

The Big Picture

The Mirondians were true to their word ... sort of. There were changes from Year Four. The first noticeable change was that the Mirondians got hard-nosed in their bargaining on contracts. They wanted more for less. Clearly, the spaceship was expensive, even for the deep-pocketed Mirondians. The hard-nosed bargaining was their way of passing on part of the cost.

The second big change was opening what the solar system community called Mirond University -- MirU for short. A campus was built at the Show Room asteroid. There solar System people learned from MSEs about advanced blip construction and management and other things that the Mirondians felt star travelers should learn. The university grew to where it taught thousands of students. When the university started, it appeared that the Mirondians finally emptied their MSE barrel staffing the university. After it got started, MSEs switched around from project to project, but very few new ones came from the mothership.

The programs taught were blip short in duration. Master's and doctorate equivalents took six months each. They were short but effective. Graduates came out with distinctly different ideas on how to get things done, and the social friction grew even more.

In month nine of Year Five, a series of demonstrations spread around the solar system community and then turned violent. The turn to violence was spooky. Earth communities have had long experience with protesting demonstrations, including those that turn violent, so on Earth, this wave was handled with average disruption. But protests and demonstrations were new to the other solar system communities. Previously, the communities had been too small, and the people of the community too familiar with each other to need them. And the idea of random property damage was sacrilege on the colonies. In the colonies, you might get angry, but you never took it out on stuff because stuff was what kept you alive. The strange actions that happened during these violent protests deeply shook the off-Earth solar communities. "Earth's bad habits are spreading out with their immigrants!" was the gossip.

The protests led to many community leaders calling for a cooling-off period. For three months, the Mirondians stopped generating new contracts, and when they started again, there were some socially protective provisions included in them.

These were stressful times, and many people were coming up with many ideas for curing the stress. In calmer times, such people are called cranks and crackpots and mostly ignored. But in these stressful times, many more people listened seriously to these strange ideas, and the speakers attracted followers.

Chapter 6: Choosing Sides

In The Trenches

Six months later, after graduating from MirU, John Foursmoots is meeting with his grant committee on his next big astronomy project. The committee is headed by Department Head Andrew Boshkov, a heavy-set man in his fifties.

Andrew says, "John, we have reviewed your proposal, and we are rejecting it."

John says, "On what grounds? This is a continuation of what I've already been working on."

"Think about it, John. Why should we fund any more big astronomy? These days, it is all covered in the Mirondian Encyclopedia."

"That is nonsense! Only one percent of what we look at with our telescopes is in Mirondian space. They don't know any more than we do about the rest of the galaxy or other galaxies. They haven't traveled there. Their knowledge should make no change to our funding."

"Sorry, John, we don't see it that way. Until we have fully digested what's in their encyclopedia on astronomy, we are holding off on the big astronomy projects."

John thinks for a moment, can think of nothing to add, and walks out. He is not happy. What he thinks is that it is time for a career change.

<<<*>>>

Not much later, John and Jason are back on Myrtle's show.

277

John says, "The Mirondian cultural systems are alien, and they are dehumanizing. We need to back off and reexamine what we humans want out of this relationship. Doing things the blip way needs to be carefully examined."

Jason says, "Doing things the blip way is increasing inequality. It isn't fair to the poor people of Earth. It isn't fair to the poor people around the solar system. If blip can't increase fairness, we need to abolish it."

Myrtle says, "I understand you have a new book out, John."

John holds up his book. "It's called *The Alienation of Mankind*. I go into detail on why Mirondian goods and technology are dehumanizing mankind." He explains more as a video runs behind him, "The highlight of the problem is the communication-intensive aspects of blip techniques. First-time viewers of blip techniques in action always leave with their jaw dropped. To first-time viewers, blip processing looks a lot like a Betty Boop animation of the 1920s in which everything is moving and everything moving is perfectly synchronized with everything else moving.

"Kids seeing a blip process in action for the first time squeal in delight. There are now several children's museums that put obsolete blip machines on display for just that reason. Getting that kind of coordination requires intensive communication. Much of the communication is machine-to-machine, but MSEs and humans are also in the loop."

He puts a communicating device on his head. It looks like an early BlackBerry. "They get into the loop using one of these. This can open up two or three communications channels simultaneously."

He gives it to Myrtle. "Give it a try."

She listens for just a moment, shakes her head, and says, "It's all gibberish to me."

"To understand this takes training. Intensive blip training. Even with that training, only some people can adapt. Many can't.

"My associates and I have done studies. Those who can adapt to this are soulless nerds. The conclusion is that humanity be being dehumanized by its adaptation to blip."

Myrtle is speechless until she announces, "John, Jason, thanks for your insights. Time for a break."

<<<*>>>

Miro-Robert has another big announcement on the Mirondian News Channel. The preliminary planning for building the next starship is complete.

Miro-Robert says, "We Mirondians have completed planning for the building of the next starship. Once again, I would like to express our gratitude to you humans. If you had not created such a wonderful solar system here, one so full of humans and resources and life in general, this project would not be possible. Our plan is to construct this new ship on your Earth's Moon. This is a good place." As he ticks off reasons, they show up as bullet items on his screen.

- "It is close to Earth, so moving the hundred thousand or so human laborers the project would take will be comparatively easy.
- "There are plenty of resources on the Moon and on nearby Earth.
- "It is not likely, but if space pirates show up, the Moon has defenses.

"Once again, we Mirondians look forward to this great leap forward in Mirondian and human cooperation."

And he signs off.

The reaction on Earth is huge and unenthusiastic. "Mirondians Plot Moon Trashing" screams a headline on the screen at Myrtle's show. John and Jason are back, and Zeb and Judy are watching the show from their offices.

John says, "To Mirondians, the Moon is just another satellite. What they think is important is that 'their baby'—the ship—should be gestated somewhere safe and as convenient as possible. But for us Earth people, the Moon is something very different. It is the unchanging goddess of the night. And I have been going through the details that the Mirondians have released. This project is huge! And the pollution it is going to create is huge! It will create a mountain—no, mountain range—of trash and pollution on the Moon's surface. Not only would the starship orbiting the Moon be an eyesore for decades, but the Moon would also develop a dusty atmosphere from all the pollution. The surface scars created would be visible to the next species. This is not a good idea."

It took a month for the Mirondians to realize that the Earthers weren't kidding. This was deeply upsetting. When they do, Miro-Robert is back on the air.

Miro-Robert says, "We have listened. We are changing the construction site to a large as yet unnamed Earth orbit-crossing asteroid. As a bonus, when we are finished, this asteroid will no longer be a threat to Earth."

Once selected, the asteroid is visited by many ships. Using blip techniques, many communities that house humans are built around it. As the project gets fully underway, the humans here number in the tens of thousands. This is by far the solar system's largest in-space city.

And the asteroid itself is being transformed by these humans, robots, and MSEs into a new Mirondian mothership, only five times bigger than the existing one. This will be a new Mirondian mothership, and when it is complete, it will carry Mirondians and humans to visit other star systems.

One of the workers on the mothership is Judy. She is living in one of the neighborhoods as the various space settlements are called by the locals. She is on a comm line with Zeb. He is on Earth. Earth and the asteroid are close by each other right now, but they are still far enough apart that there is a several second delay in the transmission due to the distance. (When they are far apart, on opposite sides of the sun, the lag gets up to sixteen minutes.)

Judy says, "Zeb, how goes it?"

Zeb says, "Doing fine here. Did you hear the news? I have a blushing bride now."

"No! Congratulations! Send me some pics and info when you get a chance."

"I'll do that. What's up?"

"I'm living and working at Mothership City now. Our company has been looking over the design of some of the HVAC components being used in eating areas, and we see room for improvement. There have been some significant changes to the surrounding components over the last month. The current fabricators are being—dare I say it?—bozos. They are still using preblip techniques. Is this something one of your groups can take over? I'm sending the specs."

Zeb takes a quick look. "Offhand, it doesn't look like anything too difficult.

"But something to keep in mind is that those people at the old place are going to be real unhappy if they lose their jobs. I'm checking the stats on that other company. This looks like this component is a big part of what they do."

"Yeah, but they are selling it to us based on its value as being artisanally crafted. This part has moved up in functional importance. It's not decorative now, and we don't need artisanal. We need fully functional and more adaptable. And we need a whole bunch more, about ten times what the previous plan called for."

"We can handle it. No problem. I'm passing your RFP on to the right group. I'm just wondering. Is there something else this group can be making for you?"

"Artisanal is their forte. Let them make something artisanal for the ground-pounders on Earth. The economy is booming. they can find something."

"Booming out in space, not so much so here on Mother Earth."

"It should be! It's a lot closer to the mothership than anywhere else in the solar system. Have those artisanals pull their heads out, and Earth can boom just as much."

Zeb just sighs at this.

<<<*>>>

John and Jason are with Myrtle again. The theme this time is a forum for those unhappy with how the Mirondian mothership project and blip are transforming Earth society.

John starts off. "This mothership project is showing us the dark side of dealing with the Mirondians. They are now squeezing human contractors, particularly those on Earth. They now want things done for a pittance, and it's do it their way or the highway. They allow no creativity or humanity in developing the product."

Jason says, "Look at the millions who are out of work because they won't conform to doing things this … this … blip way! Blip is dehumanizing. Those people doing it the blip way are turning into robots who are controlled by the Mirondians! Look at how much they have to talk with them to get anything done! Look at how often the Mirondians change their mind about what they ask for. The humans meekly say, 'Yes, sir. Whatever you say, sir.' This is not what we want for Earth! This is zombie apocalypse … with warm zombies!"

John says, "And now we are doing something about this. We are starting a group called People First! The goal of this group is to make sure that Earth stays suitable for humanity. Real humanity, not blip humanity. As our cause grows in strength and numbers, we are going to make sure that Earth stays suitable for human culture development. We want Earth to be for humans, not some weirdo aliens, or even more weirdo humans who think like aliens."

<<<*>>>

People First! starts pulling off some high profile antics. In a big wheat field in Iowa, they hack a robot harvester so it draws a crop circle that spells out "Miro Go Home" that can be seen from the air. That completed, the harvester goes back to its normal duties before any authorities show up.

Next, they hack into Myrtle's show. The hacked channel shows an imitation of Austin Powers' Dr. Evil who says, "Listen to me, you Mirondian aliens. This is your one and only warning. I will unleash upon your show room asteroid an invasion of these ..." He holds up a cute kitten. "Cute Ninja Kittens if you don't pay me ..." Finger to mouth. "One million dollars!" Then the hack ends. It is cutely executed, and it goes viral.

These first two were cute, but the next one isn't. The next protest happens at the tropical beach where we saw the Mirondians and Eggleston earlier. There are people and Mirondians walking up and down the beach and enjoying themselves.

They are interrupted by a noisy mob of protesters shouting "Mirondians, Go Home" and similar chants. The police show up and start hauling them away.

<<<*>>>

The complainers are not just on Earth. The media picks up more protesting news. Myrtle is interviewing Charles "Goodtime Charley" Monson, a man of medium build in his fifties wearing nice manager-style clothing.

Myrtle says, "Next on our show we will have the president of the Moon Boosters, Charles 'Goodtime Charley' Monson." Monson walks on and sits down. "Mr. Monson, tell us a little about yourself and why you're here."

Monson says, "I'm the head of Moon Boosters, Myrtle. Ours is an organization dedicated to building prosperity on the Moon by building business on the Moon. I'm here because, thanks to those Mirondians, it isn't happening. Instead, they are turning us into a solar system park! Christ! There was no one complaining about moon pollution or moon surface scarring when the military built MB-10 to protect us from the Mirondians. That was the biggest surface complex ever put on the Moon. Now, we can't sneeze in our spacesuits without one of these new pollution allotment licenses that Earth government thought up to stop the Mirondians. This is deeply silly! Worse, this is going to foster those Earth bureaucrats to get into even more influence peddling here on the Moon. Pfft! What little there is left to peddle. To add insult to injury, those Mirondians aren't even using the Moon as the Earth-Starship transshipment point. They built a whole new shipping complex at the Earth-Moon L-5 point. These Mirondians are taking revenge. They have turned their backs on us Moon people completely. Our brightest and best are leaving. We are soon going to be as much a backwater as Pluto!"

The Big Picture

The Moon crisis ended the Mirondian-Solar System honeymoon. Before the starship-building announcement, the solar system people viewed the trading as clearly beneficial and ignored the social stresses it was causing as a temporary, necessary evil of gaining the benefits. After the starship announcement and the following Moon crisis, solar system people began paying more attention to the tradeoffs. Some started calling them "the dark side of dealing with Mirondians."

Most of those who could work the new systems saw no tradeoffs. They saw humanity as getting better, a lot better, and they themselves were perfect examples. There were also a lot of wannabes who were enthusiastic about the new system. They weren't in it yet, but they were hopeful. These were the people taking all sorts of Earthling-sponsored training courses on how to do things the blip way. Some of these courses were authentic and very helpful, and some were scams, but in this time frame, it was hard for beginners to tell the difference.

Before the Moon crisis, a lot of humanity had no opinion. They watched, but they hadn't made up their mind one way or the other on how good all this new stuff was for humanity. After the Moon crisis, this segment of humanity steadily shrank from a large majority into the minority. People took sides.

The other side was made up of those who were unhappy and those who saw no dream coming true. Many were people who tried, but they couldn't get with the program. They just didn't think right. For them, adapting to blip was like early Industrial Age people who couldn't get used to using clocks. For those people who couldn't develop clock sense, doing shift work at a factory wasn't going to work out, and that form of the Industrial Dream was not for them.

A few of the unhappy ones were malcontents who had been in the program and then dropped out. They had been in the dream, but saw a nightmare. These were people like Foursmoots, and these people gave a lot of ammunition to those who wanted to slow the change and make it more equitable.

Many of the organizations of unhappy people started with the expressed goal of making the social change more equitable. The one that became the center of the protest movement was the People First! group with Foursmoots as its president. Foursmoots gave the movement philosophic credibility; his two second-in-commands gave it street cred and high profile. The organization became famous for its high-profile stunts and its ability to move opinions through a viral network of news seeders who kept Spacenet news sites hopping.

Work generated by the starship project was not evenly spread around the solar system. Some places got big contracts and forged ahead economically and socially when a huge influx of new people followed the contract. Other places did not and became backwaters. The Moon became a backwater, much to the dismay of Moon boosters. The Mirondians had learned their Moon lesson well. Not only did they not base the starship project on the Moon, but they also pointedly would not sign into any prime contracts that proposed to use Moon facilities.

The Moon boosters felt hugely cut out. Not only were they cut out of the starship project, but the constant acceleration transportation revolution was also passing them by. As travel became fast and cheap to everywhere in the solar system, the Moon lost its special status relative to Earth. Even before the Moon crisis of Year Five, the Moon community had started into what would become a years-long recession. Its brightest and best went elsewhere, and by Year Seven, the Moon community became a hotbed for those who chose against the Mirondians.

Chapter 7: The Year Seven Crisis

In The Trenches

Judy is watching an educational channel again, sociology this time. The professor is pontificating on the meaning of recent events.

The professor says, "The job of a government is to promote the will of the people. But ... what if the people have no will? Or many wills?

"This is the problem faced by the solar system governments of today, Year Seven since the Mirondian arrival. The many social revolutions started by the access to Mirondian products, technologies, and techniques have left many humans' heads spinning. What is right? What is okay? What is wrong? The people are trying to figure it out, and their politicians are trying even harder. The result is a whole lot of confusion. And in such times, those with strong opinions tend to win out. Sadly, the rationality of those opinions seems to matter little.

"We are in what I call The Time of Nutcases. Many fanatics are getting voted in and appointed in because they 'know'. The scary part is governments are now starting to do crazy things. This is revisiting the 1930s governments trying to solve the Great Depression."

A student communicates in a question. "Can you give us some examples?"

"Compare the governments of Titan Colony and the Moon. Titan Colony has voted in a blip-friendly slate. That legislature then voted to pay for the tuition of any Titan student who could get into MiroU. On the Moon, the Moon Boosters have been voted in. And here is an excerpt from Charley Monson's inaugural speech."

The view switches to a recording of the speech Monson made to the Moon Parliament.

Monson says, "We kicked those blip-friendly bastards out. Threw them out on their blip-happy keisters and told them to go take a ride in space!"

Back to the professor: "The next day, Moon Parliament passed a bill establishing a commission to set import and export tariffs based on the human content of the good or service being traded with the Moon. This was set up so hastily that bribery matters a lot more than human content. But it persists. It is an example of The Time of Nutcases."

The student says, "That's not even the worst, from what I hear."

The professor says, "Sadly, it's not. This week, the Moon boosters got even nuttier."

The view switches to another world-shaking pronouncement with Monson making the news again.

Monson says, "We want our fair share. And we are ready to do what it takes to get it. We are now levying a tax on transit to and from the L-5 point. It's in our orbit. That puts it in our jurisdiction. We are using the military facilities on the Moon to enforce our legislation."

The student says, "Wow! Can they get away with this?"

The professor says, "Not likely. The military on the Moon takes its orders from Earth military command. Monson is spouting pure hot air when he rattles that saber. But the media is spreading his words around. We are in The Time of Nutcases. This is something for you all to think about. This is real-time, real-world sociology in action."

<<<*>>>

Myrtle is right on top of this kind of news. Behind her are news clips of coups and countercoups taking place all over Earth.

Myrtle says, "Times are definitely unsettled here on Earth. In the six months since the Moon Boosters announced its takeover on the Moon, there has been a steady stream of coups and countercoups around the solar system, and each announcement has been met with more violence in the process.

"Some humans have called upon the Mirondians for help, and Miro-Robert has responded."

Miro-Robert is talking on the Mirondian communication channel. "We are distressed at the spreading unrest. But we see this as a passing phase. You will find your way, and we want to be there for you at the end of this painful journey."

<<<*>>>

It is now month nine of Year Seven.

Myrtle is on her show while news is on the screen behind her.

She says, "The humans aren't the only ones feeling pain now. In several locations, human kidnappers have been taking Mirondians hostage. So far, all have been released safely. But Mirondians have responded to this new threat. They are being much more cautious."

The Big Picture

In Year Seven, many parts of the solar system community plunged into anarchy. The world governments would make pronouncements, but the locals would pay no attention to them. At the local level, incumbent governments were replaced and being declared ineffectual by those who were replacing them. At first, the replacements were through legal processes (elections, recalls, and the like).

But then the coups started.

These coups were all rather independent in the beginning. Each was brought about by its own mix of issues, and there wasn't much coordination among them. But the chaos did bring intrigue, and one logical group to seek alliance with was the Mirondians.

The Mirondians steadfastly refused to get involved in local intrigues at first.

But in month nine, a new ingredient was stirred into the pot: the first kidnapping of Mirondians. Kidnapping has a long history on Earth, where wide open spaces, abandoned buildings, and hostile clans are common. But it is a rare way to treat people when you spend your entire life in a shiplike environment. The Mirondians had heard of kidnapping from their Earth history studies, but it was just another wacky Earth custom in ten thousand wacky Earth customs.

When Mirondians first came visiting, they were treated like demigods. Likewise, the first MSEs who worked with humans on blip projects were treated with the utmost respect. But familiarity bred contempt, and the chaos that the Earth communities were slipping into brought opportunity.

The first time it happened, the Mirondians complied quickly, and there was no harm done. But, like the Moon Booster coup did for political groups, the first kidnapping opened eyes for a lot of criminals. There were now ten thousand MSEs scattered around the solar system.

A dozen kidnappings followed in the next three weeks. Mirondians learned their first lessons about kidnappings very quickly.

Two things fell out of the kidnappings. First, the Mirondians and MSEs showed they could get really cautious. Second, the ransoming and antikidnapping business brought the Mirondians into dealing with local politicians in ugly, backhanded ways. They lost their virginity real quickly, and they were now taking sides along with everyone else in the solar system community.

It was a year full of changes.

Chapter 8: The End Comes

In The Trenches

It is now Year Nine, and the new ship is now about eighty percent completed.

Myrtle is on her show. The screen behind her shows a storefront on Earth for recruiting humans to go on the new mothership. It is empty of candidates.

Myrtle says, "It seems curious that while the Mirondian starship recruiting drive is in full gear now, the number of recruits seems low ... at least as best anyone can determine in these troubled times. The Mirondians say they are looking for a fifty thousand solar system people to man their ships, both of them, but the best Earth sources have been able to determine is that only ten thousand have been recruited. Here is Miro-Robert to comment on this."

Miro-Robert shows up on the big screen. He doesn't seem worried. He says, "We have done our studies of the solar system situation. We are confident that the recruiting situation will change dramatically over the next year and that we will have our people by Year Ten."

"Thank you, Miro-Robert."

Miro-Robert signs off. He is replaced on the big screen by stock videos of protesters being cracked down upon.

Myrtle says, "In Earth news, the violence of the year of coups is continuing to taper off. Order is being restored in most places."

Meanwhile, back at the Show Room asteroid, it is negotiating time. A group of military officers from Earth are ushered into the meeting room.

They are lead by General Douglas, an officer in his forties with recent spectacular successes in the crackdown phase.

In the meeting room waiting are Miro-Sam, Miro-Alex, and Miro-Bob.

Miro-Sam greets them. "General Douglas, congratulations on your military's success at restoring order."

Douglas says, "Thank you. It has not been easy."

Miro-Sam says, "And it is ... if I may say ... still a delicate situation, is it not?"

Douglas is not happy to hear this prying. "There is still much discontent, but we have a lid on it."

Miro-Alex says, "You have put many people in prison, right?"

Douglas says, "Only troublemakers ... but I admit that there have been many of those."

Miro-Bob says, "General Douglas, we have invited you here because we may be able to help you restore tranquility much faster and more reliably."

Douglas says, "What do you have in mind?"

Miro-Bob says, "You need an outlet ... a 'safety valve' for all the discontent that has built up over the last decade. I am going to propose one for you."

Douglas says, "And it is ..."

Miro-Bob says, "Have your discontented people become colonists on our ship. Empty your prisons into our ship's hold."

Douglas thinks about this and then smiles, "Yes. That will solve a lot of social unrest problems."

Miro-Alex says, "And to further help you solve your unrest issues, we will pay you handsomely for those you send."

Douglas smiles more when he hears this, "I think we will come to a mutually beneficial arrangement. Yes, indeed."

<<<*>>>

Zeb and Jason are meeting at the transport terminal again. Jason has just dropped some big news. Zeb is wide-eyed.

Zeb says, "You're doing what?"

Jason says, "Yup. I'm going on the mothership. I'm going to be a colonist."

"I admit it. You could not have surprised me more, Jason. What's your thinking behind this?"

"Earth is going to hell in a handbasket. I've tried, but I can't do anything to stop it. So, I may as well try to make a better world somewhere else. And the only way to get somewhere else is on board one of those Mirondian ships. I signed up. I'm headed for training now."

"Wow! Well, good luck at it. Let me know if I can help out."

"I will."

They get up, shake hands, and go their separate ways.

<<<*>>>

It is now Year Ten. At the mothership asteroid and city complex, much of the hustle and bustle has stopped. The building of the new mothership is complete. There are two ships here now—the old and the new Mirondian motherships. The old mothership has moved from Neptune orbit to the asteroid so some final exchanging between the two ships and the asteroid complex can be done quickly.

They both fire up and start their journeys to other stars. This would seem like the end, but there is one more twist.

Myrtle is talking on her show. "The Mirondian motherships are now headed out of the solar system. Their holds are filled with Mirodian and human colonists.

"But that is not the last of the surprises this Mirondian interlude has brought to the solar system."

The view behind Myrtle changes. It now shows a futuristic city in a thick jungle.

"About five thousand of the MSEs have decided to stay in the solar system, and they decided to stay on Earth. In another flashy display of blip-in-action, just two weeks before the motherships' departure, the Earth-loving MSEs set up a colony in the center of the Amazon jungle. How much the world has changed in just ten years! If the Mirondians had attempted this when they first landed, the whole solar system would have fought them off tooth and nail. But so many other strange things have happened that now it seems like just one more strangeness. There has been some protesting, but no violence. It is just another strangeness of the time."

The Big Picture

On the surface, Year Nine did not look a lot different from Year Eight. The anarchy continued as people with crazier and crazier ideas got their chance to try running governments. The increased craziness brought increased violence, and historians have called this period a civil war. But underneath it all, confused people were steadily getting less confused. In some places, people had worked out how to live in this new world, and they were throwing out the kooks. In their places were coming stern disciplinarians who were restoring order. The biggest problem with these new disciplinarians was that some were as crazy underneath as the kooks they were cracking down on, and they wanted glory as well as order. At first, these leaders brought serious large-scale fighting to the civil war and large-scale pogroms to the local communities.

In the case of their slow-starting recruiting drive, the ace in the hole that the Mirondians seemed to be counting on showed up in late in Year Eight, and it continued on into Year Nine. The civil wars being fought in various communities were being won by one side or the other. As the wars ended, the winners found themselves with a lot of prisoners, a lot of government debt, and a lot of discontented community members. The ones who were not happy with who won. They also found themselves meeting with Mirondian representatives who were ready, willing, and able to help them solve their discontented people problems with a special recruiting drive. Those leaders who played ball with the Mirondians found themselves with a lot fewer trouble makers and a lot of relief to their fiscal problems. Leaders were offered a plan, but many people on the street were just as discontented, even those who were on the winning side of the crackdowns. Those previously forlorn looking storefront recruiting centers got very active, and the lines in front of them consisted of people from both winning and losing sides. That is how the Mirondians filled their ships on schedule.

At the beginning of Year Ten, the two Mirondian ships departed on schedule. The old one was large, the new one was huge, and the Mirondians were very proud of both. The trading with the solar system had gone very well. There were Mirondians crewing both, and both had a mix of older MSEs and newer solar system people ready to become colonists in distant star systems.

There was one surprise wrinkle at departure time. Given the chaos and the new wealth of the Mirondians, perhaps it was inevitable, but when it happened, it was a surprise to the Earth people. About five thousand MSEs decided to stay on Earth.

The two weeks after Amazon Colony was established were filled with celebrations and farewells. With little additional fanfare, the ships engaged their drives and began their departure from the solar system. Earth's first contact with an alien civilization had drawn to a close.

Chapter 9: Epilogue

In The Trenches

This is now fifty years later. Myrtle is on her show celebrating the fiftieth anniversary of the departure. On the screen behind her, there is a scene of the motherships departing the mothership asteroid city complex and then there are scenes of little happening. The city complex is shown as mostly ghost town.

Myrtle says, "Earth and the solar system were not the same place after the Mirondian visit. The social unrest caused by Mirondian products, technologies, and blip lasted for fifty years after the departure. But the biggest difference was that mankind now had the tool for star travel. Humans had done most of the back work building the Mirondian starship, and they could do it again if they had the will. Whether to have the will was one of the topics argued hard about during the fifty years of unrest. There were a lot of people dead set against it. There were attempts to sabotage the starship-making infrastructure. There were attempts to cannibalize it to make other things. But in the end, it was decided that the solar system was such a much better place and that mankind should follow its dream."

The big screen behind her now shows a new mother ship at Mother Ship Asteroid.

"In 2134, a new starship is being dedicated. How many will follow is a question yet to be answered."

The Big Picture

An alien spaceship arriving at the solar system will bring about big changes to how the people of the solar system live.

But unlike Christopher Columbus landing on an island in the New World, the change will not be just the first step of many. Alien visits will remain isolated, one-shot events. This will be so because, using the physics we know today, star-to-star travel will always be an expensive process and one that requires years-to-centuries long journeys.

This combination of high expense and long journey times limits commerce, and commerce is what sustains large quantities of travel.

For the same reason, the number of interstellar ships that the solar system will create and send out will also be small. Without profitable commerce, the main reasons to send off ships will be to satisfy scientific curiosity and solve bitter social issues by sending off malcontents. An example of the latter is the colonists of the Mayflower trying their luck in the New World.

In sum, visits from aliens will be exciting, but rare and random events. Even when the solar system develops interstellar ship-building capabilities, the number of ships produced will be small because commerce won't be driving the activity.

Printed in the United States
By Bookmasters